Applied Condition Monitoring

Volume 8

About this Series

The book series Applied Condition Monitoring publishes the latest research and developments in the field of condition monitoring, with a special focus on industrial applications. It covers both theoretical and experimental approaches, as well as a range of monitoring conditioning techniques and new trends and challenges in the field. Topics of interest include, but are not limited to: vibration measurement and analysis; infrared thermography; oil analysis and tribology; acoustic emissions and ultrasonics; and motor current analysis. Books published in the series deal with root cause analysis, failure and degradation scenarios, proactive and predictive techniques, and many other aspects related to condition monitoring. Applications concern different industrial sectors: automotive engineering, power engineering, civil engineering, geoengineering, bioengineering, etc. The series publishes monographs, edited books, and selected conference proceedings, as well as textbooks for advanced students.

More information about this series at http://www.springer.com/series/13418

Nikolai K. Myshkin · Liubou V. Markova

On-line Condition Monitoring in Industrial Lubrication and Tribology

Springer

Nikolai K. Myshkin
Tribology Department, V.A. Belyi Metal-Polymer Research Institute
Belarus National Academy of Sciences
Gomel
Belarus

Liubou V. Markova
Belarus National Technical University
Minsk
Belarus

ISSN 2363-698X ISSN 2363-6998 (electronic)
Applied Condition Monitoring
ISBN 978-3-319-61133-4 ISBN 978-3-319-61134-1 (eBook)
DOI 10.1007/978-3-319-61134-1

Library of Congress Control Number: 2017945713

Printed on acid-free paper

This Springer imprint is published by Springer Nature
The registered company is Springer International Publishing AG
The registered company address is: Gewerbestrasse 11, 6330 Cham, Switzerland

Preface

The year 2016 has become a jubilee of tribology connected to presentation of Prof. Peter Jost report to the British government in 1966. In this report, Prof. Jost has placed emphasis on the economic importance of the problem of friction and proposed the formulation of the scientific concept as “tribology” (from the Greek word “tribos”—rubbing): the science and technology of interacting surfaces in relative motion and of related subjects and practices. It was a reflection of a new synthetic approach to the problem having millennial roots. Engineering solutions of friction and lubrication problems by the ancient civilizations have led to the appearance of ski and wheel. Creation of the first lubricants based on vegetable and animal products have made possible the technological progress of the mankind.

Nowadays, friction is responsible for crucial problem of engineering—wear of machines and mechanisms. The expenses due to rehabilitation of the worn-out machine parts are enormous, whereas increase of their lifetime is very profitable. Drop of the losses on friction and energy saving, wear reduction in machines and mechanisms together with economy in materials, elimination of harmful ejections, and other topics are now a part of the tribology subject. In Jost’s opinion, in the course of its half-centennial development, the notion of tribology has penetrated into all the fields of human activities and natural phenomena connected with friction processes.

Modern tribology can be divided into the main areas specified as: fundamentals—contact mechanics, surface physics, and chemistry; materials science—development of materials, coatings, and lubricants for tribosystems; technologies—surface engineering; design—development of the efficient tribosystems; tribodiagnostics—methods and devices for condition and wear monitoring; testing, data processing and presentation in the form of standards, recommendations, as well as distribution of tribology achievements to research and engineering community.

This book covers the area of tribodiagnostics which is still not extensively presented in the technical literature. Its main emphasis is done to the problems of monitoring the lubricant, as it is an essential component of a tribosystem. During operation, the tribosystem is exposed continuously to the environment, elevated temperatures, speeds, and loads inducing variations of the chemical and physical properties of the lubricating oil. The main causes of impairment of oil properties are modification of the oil chemical structure (oxidation, thermal destruction, decomposition of additives), higher content of water, and chemical impurities in the oil. Lubricating oil condition reflects the processes evolving in the tribosystem as a whole. On-line and in-line oil analysis systems provide timely data on the need for routine service or early warning of oncoming catastrophic events. The authors have presented an extensive review of the analytical methods, calculations related to the physics of the techniques used in their research, as well as the principles and designs of the monitoring devices, including their own ones.

The content of the book includes the introduction, six chapters covering the main aspects of the lubricating oil monitoring, and concluding remarks. The introduction positions the place of condition monitoring and predictive maintenance in the whole subject of tribology. First four chapters are related to the methods and tools for evaluation of oil physicochemical properties, as well as very important problems of monitoring the oil viscosity, water content, and soot concentration in oil. Fifth and sixth chapters cover the wear prediction for tribosystems based on debris analysis and trends in condition monitoring for tribology. Conclusions formulate the prospects of development of condition monitoring, as well as the main problems arising on this way.

The whole multidisciplinary subject of tribology is widening quickly and covers new areas of human activities. Nowadays, it is a great area of research and development having key validity for engineering in industry, transportation, agriculture, and even medicine. Condition and wear monitoring is an important part of tribology, so the authors hope that the current book can be useful for researchers and engineers working in such an important and useful area.

The personal experience of the authors is mainly connected to their work in the Metal-Polymer Research Institute of the Belarus National Academy of Sciences, as well as to the international cooperation started as early as 1980s with the Swansea Tribology Center in Great Britain, then continued with the Korea Institute of Science and Technology, and Austrian Center for Competence in Tribology. The authors are very much grateful to their colleagues in cooperation, especially to Prof. O.K. Kwon, Dr. Hosung Kong, Dr. H.-G. Han, and Dr. E.-S. Yoon (KIST). Their sincere thanks should be given to all the colleagues in MPRI, especially to M.S. Semenyuk and V.M. Makarenko.

Gomel, Belarus Nikolai K. Myshkin
Minsk, Belarus Liubou V. Markova

Contents

Nomenclature

Chapter 1

f_1, f_2	Oscillating circuit frequencies;
$U(f_1)$, $U(f_2)$	Voltages at output of two oscillating circuits;
$U(\varepsilon), U(h), U(T)$	Signals corresponding to dielectric permeability of oil, oil level, and oil temperature, respectively;
$\alpha_{\lambda 1}, \alpha_{\lambda 2}$	Absorption coefficients at wavelengths λ_1 and λ_2, respectively;
$\Delta\alpha$	Difference in absorption coefficients;
P_0 and P	Power of exciting radiation and fluorescence, respectively;
C	Fluorophore concentration;
C_q	The concentration of the quenching species;
Q	Quantum yield;
β	Molar absorptivity of fluorescence component;
l	Length of optical path of radiation in substance under testing;
k	Stern-Volmer quenching constant;
F	Fluorescence emission ratio;
I_0, I	Intensity of fluorescence without a quencher and with a quencher, respectively;
$I(\lambda)$	Spectral fluorescence intensity of oil;
$I_{eff}(\lambda)$	Efficient spectral fluorescence intensity;
$I^*(\lambda)$	Relative spectral fluorescence intensity of oil, $I^*(\lambda) = I(\lambda)/I_{\max}$;
$I^*_{fr}(\lambda), I^*_A(\lambda), I^*_B(\lambda)$	Relative spectral fluorescence intensity of fresh oil and two samples (*A* and *B*), respectively;
$I_{\Delta\lambda l}, I_{\Delta\lambda sh}$	Fluorescence intensity measured in long-wave range and in short-wave range, respectively;
$\Delta\lambda_l, \Delta\lambda_{sh}$	Long-wave and short-wave ranges, respectively;
$\Delta\lambda_R, \Delta\lambda_G, \Delta\lambda_B$	Red, green, and blue ranges, respectively;

$S(\lambda)$	Spectral sensitivity of the photodiode;
$S_R^*(\lambda), S_G^*(\lambda), S_B^*(\lambda)$	Relative spectral sensitivity of RGB photodiode in red, green, and blue ranges, respectively;
$I_{\max}$	Maximal spectral intensity of oil fluorescence;
$S_{\max}$	Maximal spectral sensitivity of photodiode;
$S^*(\lambda)$	Relative spectral sensitivity of photodiode, $S^*(\lambda) = S(\lambda)/S_{\max}$;
J_R, J_G, J_B	Output signals of RGB photodiode in red, green, and blue ranges, respectively.

Chapter 2

ν, η	Kinematic and dynamic viscosity of liquid, respectively;
C	Viscometer constant;
t	Mean arithmetic time of outflow;
ω_r	Rotor speed;
M	Torque;
K	Device constant of the rotational viscometer;
τ	Shear stress;
D	Shear rate gradient;
ρ_l, ρ_b, ρ_w	Densities of the test liquid, ball material, and waveguide material, respectively;
R_b, v_b	Radius and velocity of ball, respectively;
g	Gravitational acceleration;
A, f	Amplitude and oscillation frequency, respectively;
V, ω	Velocity and circular frequency of the acoustic wave, respectively;
ω_a, ω_l	Circular frequency of natural oscillations of the plate in air and liquid, respectively;
φ	The phase shift;
Z_w, Z_l	Wave resistance (characteristic impedance) of the waveguide and liquid materials, respectively;
μ	Elastic modulus of the waveguide on shear;
a.v.	Acoustic viscosity;
ρ_{pl}, E_{pl}, σ_{pl}	Density, Young's modulus, and Poisson's ratio of the plate material, respectively;
d_{pl}	Thickness of the plate;
u_y	Component of the vector of the displacement of plate particles along the Y-axis;
v	Velocity of the liquid;
$\vec{H}_f$	Magnetic field strength;
H	Strength of the stationary bias magnetic field;
$\vec{F}_y$	Friction force;

δ	Depth of the penetration of excitation induced by plate oscillations into the liquid;
f_a, f_l	Frequency of oscillations of the plate in air and liquid, respectively;
$f_{l,min}$, $f_{l,max}$	Minimum and maximum measured frequency of oscillations of the plate in the liquid, respectively;
f_{ex}	Frequency of the exciting current;
ξ	Viscosity damping factor;
λ	Magnetostriction;
η_{min}, η_{max}	Minimum and maximum measured dynamic viscosity of the liquid, respectively;
t_d	Duration of oscillation decay;
Y_0	Specified limiting value of oscillation amplitude;
T	Oil temperature;
M_w	Molecular weight;
C_m	Concentration of VI improver.

Chapter 3

C_w	Water content in oil;
C_s	Level of oil saturation with water;
C_t	Reagent concentration in titrant;
P	Partial pressure of water vapors in oil;
P_s	Partial pressure of water vapors in oil at saturation;
RS	Relative oil saturation with water;
aw	Water activity;
V	Volume of solution under analysis in titration;
V_t	Titrant volume;
I_R, I_C, I_L	Resistive, reactive, and leakage current, respectively;
T	Temperature;
I_p	Peak current value;
R	Resistance of the series resistor;
C	Capacitance of the capacitor;
τ	Time constant, $\tau = RC$.

Chapter 4

δ	Adsorption layer thickness;
D	Diameter of a particle;
E_v	Energy of attraction between particles;
k	Boltzmann's constant;
T	Temperature;
C	Capacity;
γ	Total electric conductivity;
ρ	Specific resistance;

ε	Oil permittivity;
S	Surface area of capacitor plate;
d	Gap between plates;
g	Active cell conductivity;
f	Frequency;
c	Soot concentration;
λ	Wavelength of optical radiation;
$K_{ext,\lambda}$	Extinction coefficient at wavelength λ;
l	Optical path length in oil;
τ	Optical transmission;
h	Thickness of layer into which radiation penetrates;
n_1, n_2, n_0	Refractive index of optical material, oil under testing, and medium (air), respectively;
n_D	Refractive index of optical glass measured at wavelength of 589 nm;
θ_1	Angle of incidence of optical radiation at interface;
θ_{1cr}	Critical angle of incidence;
θ_{1max}	Limit angles of aperture angle in the light guide;
$\bar{n}_2, \chi$	Complex refractive index and absorption index of oil;
U_{out}	Output electrical voltage;
φ_{ap}	Aperture angle of radiation source;
α_λ	Coefficient of absorption;
$R(\theta)$	Reflectivity of interface between two mediums;
m	Number of reflections;
P_{in}, P_{ref}, P_{refr}, P_{inc} and P_{out}	Input, reflected, refracted, incident, and output powers, respectively;
$P^{(1)}_{ref\perp}, P^{(1)}_{inc\perp}(\theta_1, n_2)$, $R_\perp(\theta_1, n_2)$	Power of the reflected radiation from the light guide—external medium interface, power of incident radiation, and reflectivity of light guide—external medium interface when polarization plane of ray is perpendicular to incident plane, respectively;
$P^{(1)}_{ref\parallel}, P^{(1)}_{inc\parallel}(\theta_1, n_2)$, $R_\parallel(\theta_1, n_2)$	Power of the reflected radiation from the light guide—external medium interface, power of incident radiation, and reflectivity of light guide—external medium interface when polarization plane of the ray is parallel to the incident plane, respectively;
P_{max}, P_{min}	Maximal and minimal values of output power, respectively;
M	Modulation index;
t, L	Thickness and length of the plate;
D_r, L_r	Diameter and length of optical rod;

$\Delta U/U$	Relative change in the output signal U;
E_v, E_s	Attractive and the steric repulsion energy between two electrical point dipoles, respectively;
s	Distance between the surfaces of spheres;
H	Hamaker's constant;
N	Concentration of adsorbed molecules on the surface;
A	Number of moles or grams of adsorbate per unit surface or per unit mass of the adsorbent;
G	Gibbs adsorption;
c_a	Equilibrium concentration of the adsorbate;
σ	Surface tension;
G_i, μ_i, c_i, c_{0i}	Gibbs adsorption, chemical potential, equilibrium bulk concentration, and initial concentration of *i*th substance, respectively;
A_i	Adsorption of *i*th substance;
N_{total}	Total number of particles *i*th substance in the system;
V, S_q	Phase volume and square of interface, respectively;
h_s, V_s	Thickness and volume of the surface layer, respectively;
τ_a	Time of adsorption;
k_a	Constant of adsorption rate;
Q	Degree of surface filling by adsorbate;
Q_{eq}, A_{eq}	Degree of surface filling by adsorbate and adsorption at equilibrium, respectively;
F, F_{ad}, F_r, F_c, F_{sep}	Molecular, adhesion, drag, lifting, and separating force, respectively;
R	Radius of the sphere;
h_{sp}	Distance between the sphere and plate;
P	Particle weight;
η	Friction coefficient;
f	Force acting on the dipolar charge;
E	Electrical field intensity;
q, l_d, p	Charge, length, and moment of the dipole, respectively.

Chapter 5

Δm	Mass loss;
η	Dynamic viscosity of oil;
a, V	And volume of particle, respectively;
$\vec{H}$, H_z, H_r	Magnetic field intensity and its components along the z-axis and radial directions, respectively;
$\vec{\mu}$, $\vec{\mu}_i$, $\vec{\mu}_j$	Particle magnetic moment, magnetic moment of *i*th and *j*th particle, respectively;

I_s	Saturation magnetization of magnetic material;
ρ_p, ρ	Densities of particle and medium, respectively;
g	Acceleration of gravity;
W, W_r	Wear and wear rate index, respectively;
D_L, D_S	Optical densities of the precipitate at the beginning and end of the ferrogram;
ε	Relative permittivity;
N_0	Particle demagnetizing field coefficient;
χ	Magnetic susceptibility of the monolithic material;
$\vec{r}_{ij}$	Distance between *i*th and *j*th particles;
U_M	Energy of dipole–dipole interaction;
θ_i, θ_j	Angles between $\vec{\mu}_i$ and $\vec{r}_{ij}$, $\vec{\mu}_j$ and $\vec{r}_{ij}$, respectively;
$U_{br.r}, U_{br.t}$	Energies of rotational and translational motion, respectively;
k	Boltzmann's constant;
T	Temperature;
n	Number of particles per unit volume;
N_{ad}	Surface concentration of adsorbed molecules;
δ	Thickness of adsorption layer;
s	Distance between surfaces of particles;
A	Hamaker's constant;
U_V	Energy of intermolecular interaction for two spherical particles;
N_c	Average number of particles in chain;
B	Second virial coefficient;
L	Distance between the chains;
I_{01}, I_{02}	Intensities of incident and transmitted radiation, respectively;
k_λ	Absorption coefficient of medium for wavelength equal λ;
l	Length of optical path in medium;
Φ	Flux of optical radiation;
A_s	Area of irradiated surface;
U	Output voltage of photoreceiver;
U', U_0, U_1	Output signals when the cell is empty, cell with fresh oil, and cell with used oil, respectively;
U_2	Output voltage when magnetic field affects the oil;
S	Integrated sensitivity of photoreceiver;
D_1, D_2	Optical density of fresh and used oil, respectively;
C_t	Total oil contamination index;
τ	Transmittance of the optical fiber;
E	Illumination;
R	Reflectance of substrate;
Δt	Time interval;

n	The number of measuring systems;
N_{ji}, W_{rj}, W_{sj}	Indicator of particle concentration measured in the *i*th zone, wear rate index, and wear intensity index of *j*th measurement, respectively;
$k(\tau_p)$	Coefficient of the transfer;
F	Proportionality coefficient;
τ_p	Pulse duration;
L_{coil}	Length of coil channel;
v_p, v_{oil}	Particle velocity and average oil velocity, respectively;
f	Signal frequency.

Chapter 6

Δ_{CR}	Chemical destruction indicator;
$\Delta\lambda_R, \Delta\lambda_G, \Delta\lambda_B$	Red, green, and blue waveband, respectively;
CR	Chromatic ratio;
$U_{R,fresh}$, $U_{R,used}$	Output signals in the red waveband during analysis of the fresh and used oils, respectively;
$U_{G,fresh}$, $U_{G,used}$	Output signals in the green waveband during analysis of the fresh and used oils, respectively;
$\Delta D_R, D_G, \Delta D_B$	Changes of the oil optical density in red, green, and blue wavebands, respectively;
$D_{R,fresh}$, $D_{G,fresh}$, $D_{B,fresh}$	Optical densities of the fresh oil in the red, green, and blue bands, respectively;
$D_{R,used}$, $D_{G,used}$, $D_{B,used}$	Optical densities of the used oil in the red, green, and blue bands, respectively;
RS	Relative saturation;
$a.v.$, Vis	Acoustic and kinematic viscosities of oil, respectively;
ρ	Oil density;
Q	Output linguistic (fuzzy) set;
Q_1, Q_2, Q_3	Values of the linguistic variable good, satisfactory, and bad, respectively;
W	Weighting coefficient of the rule;
X, A, B, C, D, E	Input linguistic (fuzzy) sets;
X_j, A_j, B_j, C_j, D_j, E_j	Values of the input linguistic variables input membership functions;
$\mu_{Aj}(Dr), \mu_{Bj}(Dg), \mu_{Cj}(Db), \mu_{Dj}(CR), \mu_{Ej}(Vis)$	Input membership functions;
$\bar{\mu}_{Aj}(Dr), \bar{\mu}_{Bj}(Dg), \bar{\mu}_{Cj}(Db), \bar{\mu}_{Dj}(CR), \bar{\mu}_{Ej}(Vis)$	Numerical values of the input membership functions;

σ, c	Mean square deviation and the mathematical deviation of the Gaussian membership function, respectively;
x_i	Input variables
$\bar{\mu}_k$	Values of the degree of truth (cutoff level) for the premises of the kth rule;
$\bar{\mu}'_k$	Degree of truth of the consequents of the kth rule;
q, q_1, q_2	Output numerical variable and its critical values, respectively;
$\mu_{Qj}(q)$	Output membership function of the jth linguistic value of the output variable q;
a_j, c_j	Parameters of the triangular membership function;
$\mu_1(q), \mu_2(q), \mu_3(q), \mu_4(q)$	Cutoff output membership functions for rules 1, 2, 3, and 4, respectively;
$\mu_{\sum}(q)$	Membership function of the set union;
q_0	Integral numerical oil state factor.

Introduction

Progress in mechanical engineering is based on developing the complexity of machines, widening their operation temperature, increasing velocity, and contact pressure. The reliability of tribosystems (bearings, guides, gears, seals, etc.) should be also increased. The new positioning drives with digital control for aerospace and precise manufacturing, etc., were developed, not mentioning the MEMS and NEMS. These factors require development of predictive maintenance based on the estimate of the real state of machinery. Condition wear monitoring is an important part of maintenance preventing the failures. Its main target is an estimate of working performance of tribosystems, forecasting the probable failures, and predictive maintenance for long-term operation with the optimum performance. Vibration analysis, analysis of wear debris presence, and oil working ability are widely used techniques of tribodiagnostics. Oil condition monitoring often can detect the disorder in normal wearing of the tribosystem earlier than the analysis of vibrations and accelerations.

The present monograph considers the methods of diagnostics used in on-line condition monitoring of lubricated tribosystems. The review of the existing methods and description of the new ones developed by the authors are presented. Case studies illustrating the applications of the diagnostic methods and means developed by the authors are described.

Complex machines often comprise hundreds and even thousands of friction units. The failure of one of them can result in the malfunction of the machine as a whole, its break and failure. Failure prevention in machines, devices, and structures is becoming more and more urgent. As the society employs high technologies, machine designs is getting complicated and the elimination of consequences of failures and in some cases catastrophes is becoming very expensive (e.g., airplane accidents).

Presently, terms of maintenance are commonly set on three principles: 1—maintenance after the appearance of defect in the machine; 2—maintenance by specified schedule; 3—maintenance according to the real state of the machine.

The first type of maintenance is the cheapest, but it is permissible only for not important machines. The second principle is most widely used nowadays.

Maintenance is performed in line with recommendations of the producer by a time schedule composed on the basis of norms irrespectively of the real state of machines. It is called preventive maintenance. If the periodicity of maintenance is determined by statistic analysis, the overhaul period is commonly a time during which $\geq 98\%$ of devices operate without failures. But it turns out that $\geq 50\%$ of all preventive maintenances are performed needlessly. Moreover, for many machines, preventive maintenance and repair do not result in less frequent failures.

The third principle of maintenance is the most progressive since it provides the maximal life of equipment with the minimal expenses. In this case, the maintenance is performed only when it is required due to a high probability of failure.

Recently, real predictive maintenance is using more often than preventive one. This is confirmed by results of the analysis of different maintenance types for roller bearings. If one uses the time schedule, then as the wear of a friction unit deviates from the normal wear, the unit can fail before maintenance or the serviceable unit is maintained. When new equipment is used for which the schedule is not set, yet it is also advisable to carry out maintenance based on the diagnostics of its current condition.

In this connection, the diagnostics of lubricated moving joints is nowadays one of the most important factors of increasing the equipment reliability.

The principle of maintenance based on the real condition of a mechanism involves the methodology comprising a diagnostic system, wear prediction, and maintenance according to the conclusion on the mechanism serviceability (Fig. 1).

For diagnostic, it is of primary importance to find the proper characteristics of the tribosystem that reflect wear symptoms. Then, the sensors should be selected

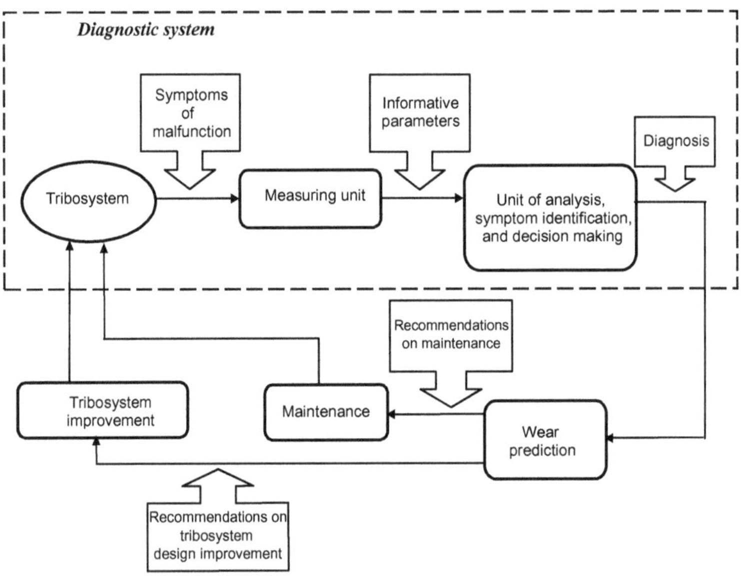

Fig. 1 Role of diagnostic system in provision of tribosystem reliability

registering reliable information. The measured characteristics should be analyzed, and the technical condition of the tribosystem is determined on the basis of a priori knowledge. The decision is commonly made on the basis of optimal decision-making.

The development of methods and instrumentation for estimating the tribosystem condition is the primary task in the creation of diagnostic systems. The methods based on vibration registration, acoustic emission methods, electrophysical methods, and temperature ones are the basics. Also other methods are applied based on the oil analysis founding the wear debris and oil performance characteristics. Vibration analysis is mostly used in industrial equipment, aircraft turbines, sea craft machines, and helicopter and jet plane engines. The analysis of oil provides data on both the oil condition and the wear of the mechanism. Typically, oil analysis is applied to monitor the condition of internal combustion engines, jet turbines, and other types of engines. Depending on the type of a machine to be diagnosed, sensors in the measuring unit can differ significantly in their complexity.

Anyway the experience shows that the monitoring should be based on the combination of different methods and devices, especially in case of important types of machinery, because the price of failure can be enormously high. The examples of such failures are numerous in such areas as space exploration, aviation, navy, heavy industrial, and transportation machinery.

Chapter 1
Methods and Instruments for Condition Monitoring of Lubricants

1.1 Introduction

Lubricant is a very important component of tribosystems, so its condition and degradation are important factors of efficiency and state of the whole machine. In using oils we face a problem of determining the replacement moment. Most motor oils are nowadays replaced on the basis of operating time intervals, which are recommended by the producers of vehicles. However, the real service life depends essentially on the oil quality, engine type, operation conditions, and maintenance of the machine.

Lubricating oils are divided by mineral, synthetic, vegetable, and animal ones. Mineral oils produced by refining crude oil have a base consisting of paraffinic, naphthenic, aromatic, and naphthenic-aromatic hydrocarbons. Synthetic oils comprise the mix of low-molecular and high-molecular hydrocarbon molecules. Synthetic oils in contrast to the mineral ones hold high working temperatures without decay and evaporating and stay movable at low temperatures. However, a relatively high cost of synthetic oils impedes their wide application. Along with purely mineral or synthetic oils their mixes are used ever more frequently, viz. so-called half-synthetic (or partially synthetic) oils.

Presently the requirements to the service characteristics of oils have become more stringent. That is why certain special additives are introduced to the base oil, viz. dispersing, antioxidation, antifriction, antiwear, antiscoring, thickening, anticorrosion, or antifoaming ones.

In tribosystem operation the oil experiences a considerable oxidative aging caused by high temperature and interaction with nitrogen oxides, moisture, and air. In addition, the availability of additives decreases or they are destructed thus resulting in deterioration of oil efficiency. Thus, in depleting of antioxidation additives the oil loses its capacity to resist oxidation. This fact results in increasing oil viscosity and acid formation that causes corrosion of metallic parts. The application of such oil leads to intensive wear and failure of mechanisms.

N.K. Myshkin and L.V. Markova, *On-line Condition Monitoring in Industrial Lubrication and Tribology*, Applied Condition Monitoring 8,
DOI 10.1007/978-3-319-61134-1_1

Methods based on analysis of wear debris and contamination particles are widely used now. They allow understanding the physical processes in the oil but not the nature of chemical transformations in it. In this connection it becomes urgent to develop methods and means for evaluating the oil condition, modification of its chemical composition, and occurrence of new chemical products.

1.2 Laboratory Methods of Oil Monitoring

Service characteristics of lubricating oils are evaluated by quality characteristics. To determine them standard methods have been developed including the measurement of acid and base numbers, kinematic viscosity, flash temperature, etc. A brief description of these methods is presented in Table 1.1 [1].

In addition to the standard methods, a wide spectrum of laboratory methods for lubricant analysis is used, viz. from electrochemical analysis of hydrolysis products to mass spectroscopy of hydrocarbon structure. Qualitative analysis allows one to determine oil chemical composition, insoluble component content, and state of additives.

Laboratory methods are commonly highly informative, sensible, and accurate. However they are time-consuming and analytical data are delayed in time. Moreover, when taking an oil sample its representativeness is somewhat problematic. Also a high cost of laboratory methods impedes their wide usage. It is evident that for optimal use of lubricant, its timely replacement and provision of reliable operation of mechanisms one should apply methods and instruments for

Table 1.1 Some standard methods for oil analysis

Oil quality characteristic	Description of method	Goal of analysis
Total acid number, mg KOH/g (ASTM D664, ISO 6619—88)	Potentiometric titration of oil sample with alcohol solution of potassium hydroxide	Oil aging from oxidation is assessed
Total alkali number, mg KOH/g (ASTM D4739, ISO 6619—88)	Potentiometric titration of oil sample with alcohol solution of hydrochloric acid (HCl)	Oil capacity to neutralize acids
Kinematic viscosity, mm^2/s (ASTM D445/446, ISO 3104—94)	Measurement of the period during which a certain oil volume flows away by gravity at constant temperature (40 or 100 °C) with a calibrated glass viscometer	Viscosity as one of the basic oil characteristics is assessed. Decreasing viscosity evidences the contamination of the system with fuel or coolant liquid
Flash temperature, °C (ASTM D92)	The oil sample is heated at a fixed rate up the flash of the oil vapors (flash temperature) in an open crucible above the oil surface from an ignition device	Availability of light-volatile and ignitable materials is assessed. For example, low ignition point of the motor oil evidences the fuel content of the oil

continuous condition monitoring of oil. They should operate in real time and combine low cost of analysis and high validity.

1.3 Methods and Means for Real-Time Monitoring Lubricant Performance

The sensors discussed here and used in devices for diagnosing the oil quality are based on measuring the following characteristics: viscosity, dielectric permeability, conductivity, corrosive activity, acid and base numbers, spectral absorption, optical density, oil fluorescence, etc. These characteristics account for the degree of oil oxidation, its contamination by fuel, coolant fluid, water and products of additive decomposition (Table 1.2).

Viscosity sensors. In oil oxidation carbonic acids are dimerized, average density of molecules increases resulting in increase in viscosity. Viscosity sensors are based on measuring the time of acoustic radiation passage, phase shear, or variation in the resonance frequency of a quartz generator. These methods allow us to develop solid–state sensors produced by chip technology.

Sensor produced by the chip technology is described elsewhere [2]. Figure 1.1 shows a typical viscosity sensor, consisting of silicon base containing protruding nickel-coated finger elements. The voltage of a specified frequency is fed to the finger elements from an oscillator. It is designed for measuring the power required to maintain forced oscillations at a specified frequency. As the oil viscosity varies the damping force acting on the finger elements put in the oil varies leading to varying power required to maintain the forced oscillations. Therefore, the variation in the required power corresponds to the variation in the oil viscosity. When evaluating viscosity, the oil temperature, measured by a temperature detector with sensitive element *1* attached to the same substrate, is taken into account. The measurement control and data processing are performed by a microprocessor.

The limited usage of these sensors is caused by two main factors. One is that the oil viscosity increases in oxidation and, simultaneously, decreases due to the fuel contaminating the motor oil. The second reason is insufficient reliability and validity of these sensors.

Permittivity sensors. In operation of mechanisms, the permittivity of the oil increases as a result of thermal oxidation. In addition, the permittivity of the oil varies with contamination by coolant fluid, fuel, water, and product of partial fuel combustion (soot). Water, which is in oils, is the most wide-spread liquid contaminant. Dielectric constants of these contaminants are much higher than that of pure oils; therefore, their availability makes the permittivity essentially higher. Therefore, increasing permittivity of the oil points to the deterioration of its quality and can be used as a diagnostic parameter.

Presently a number of devices have been developed on the basis of sensors whose capacity depends on the permittivity of the oil [3–6].

Normally the sensor is a capacitor formed by coaxial cylindrical parts. Thus, in Fig. 1.2a capacitive sensor is shown for device evaluating the aging [5]. It

Table 1.2 Characteristics of oil quality used in prompt diagnosis of oil efficiency

The reason for oil quality deterioration	Oil quality characteristics								
	Viscosity	Dielectric permeability	Conductivity	Corrosive activity	Acid number	Base number	Spectral absorption	Optical density	Fluorescence
Oil oxidation (carbonic acids and complex ethers)									
Oil contamination with fuel and coolant liquid									
Worked out additives									
Contamination with water									

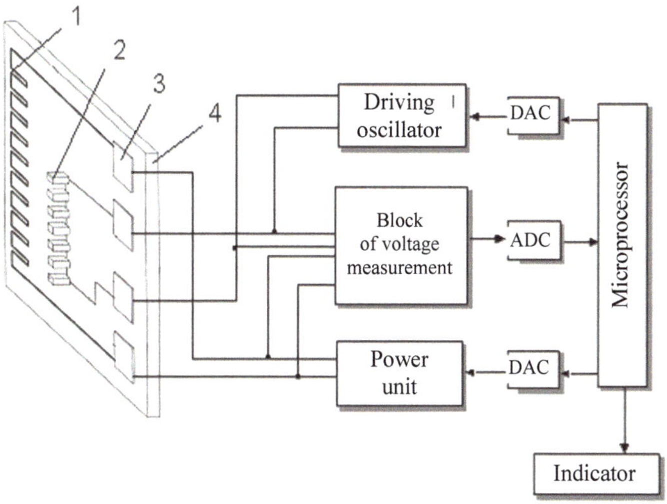

Fig. 1.1 Viscosity sensor: *1* sensitive element of the temperature detector; *2* nickel-coated finger elements; *3* electrical contacts; *4* silicon substrate; *DAT and ADT* digital-analog and analog-digital converters [2]

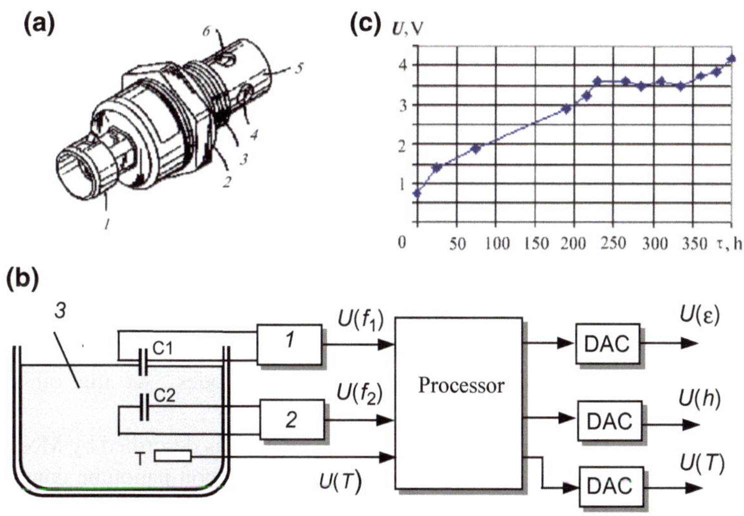

Fig. 1.2 Device for monitoring oil aging based on measuring dielectric permeability of oil: **a** sensor [5]: *1* electric connector; *2* casing; *3* thread; *4* internal conducting electrode; *5* external screening electrode; 6 through holes; **c** dependence of the output signal proportional to the dielectric permeability of oil on the operation time; **b** device schematic; C1 and C2 compensating and measuring capacitors; *1* and *2* oscillating circuits; DAC—digital-analogous converters

comprises housing *1* with external screening electrode and internal conducting electrode *2* that are connected to electric connector *3*. The housing has a hole in the side and holes *4* to pump the oil through the space between the external and internal electrodes. The sensor is mounted in the case and connected to the electronic block of the device.

To exclude the dependence of the output signal on the oil level in the case (degree of the sensor filling with oil), type of oil, and temperature effect on the capacitor parameters, two capacitors are used (Fig. 1.2b), viz. a measuring (C2) and a compensation (C1) one. The capacitors are the elements of the oscillatory circuits (1 and 2) whose frequencies f_1 and f_2 are in reverse proportion to the capacitances of the respective capacitors. To take into account the temperature dependence of permittivity of the oil the temperature meter is installed in the sensor. The signals of the oscillatory circuits $U(f_1)$ and $U(f_2)$, and temperature meter are fed to the microprocessor or microcontroller, which computes the permittivity and level of the oil. The computed values are transformed from the digital to analog form by DAC. The signals corresponding to the permittivity of the oil $U(\varepsilon)$, oil level $U(h)$ in case *3*, and oil temperature $U(T)$ are shown in the device indicator. Figure 1.2c presents the dependence of the output signal, proportional to the permittivity of the oil, on the operation time.

The diagnosis devices based on the measurement of permittivity are the most popular owing to their relatively low cost, sufficient sensitivity and simplicity of installation in the lubrication system.

Nowadays devices Kavlico Oil Quality Sensor (Kavlico Co.) and Lubrigard Oil Condition sensor Dielectric Sensor (Lubrigard Co.) are available in the market. The devices give integral characteristic of the oil quality. However, they cannot distinguish between the contribution of oil oxidation and the effect of water, fuel and coolant fluid content.

Sensors of oil conductivity. The diagnosis of the oil quality by its electrical conductivity is based on the fact that such impurities as sulfuric acid and water have a considerably higher conductivity than that of the oil.

Hedges et al. developed a conductivity sensor based on polymeric bead matrix technology [7]. The basic element of the sensor is an electrochemical cell including the polymeric bead matrix installed between electrodes. As the oil polarity increases, the conductivity across the matrix increases.

A conductivity sensor with carbon nanotube element is described by Moon, et al. [8]. The sensor technique is based on correlation of carbon nanotube conductivity with TAN of the test oil.

The conductivity sensors gives information on oil quality based on the acid and water content in the oil. However, it does not take into account the other contaminants. Besides, shot lifetime of the sensing elements limits the on-line sensor application.

Sensors of corrosive activity. Oil quality is assessed by its corrosive activity. The sensor corrosion is a direct result of additives aging and accumulation of contaminants in the oil.

The sensitive element of the sensor [9] consists of two resistive elements R1 and R2 mounted to the common substrate *1* (Fig. 1.3a) between the contact areas *2*, *3*,

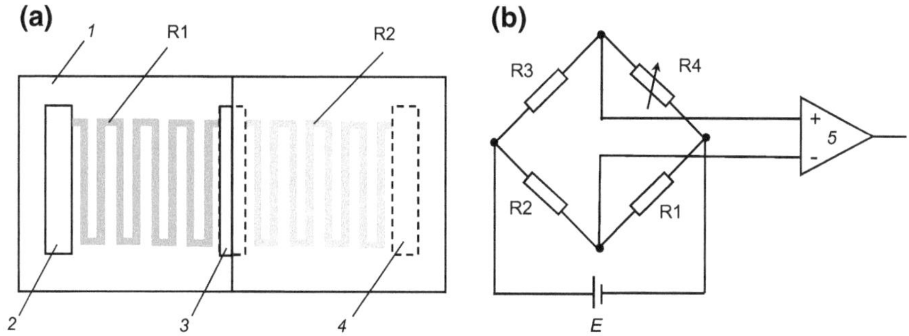

Fig. 1.3 Sensor of oil corrosive activity [9]: **a** sensitive element: *1* substrate; *R1 and R2* resisting elements; **b** sensor's electric scheme: *R1 and R2* resistors; *5* differential amplifier

and *4*. One of the resistive elements (R1) is put in the oil while the second one (R2) is isolated (sealed) from the oil.

Change of resistance in R1 due to corrosion shows the content of corrosion contaminant. It is measured by bridge (Fig. 1.3b). The disadvantage of this sensor results from necessity to change the sensitive element at oil replacement.

Acid number sensors. The number of acids in oil increases gradually because of oil thermal oxidation and accumulation of NO_x, SO_x, etc. in operation, which yields an increase in the acid number. The standard method for determining the acid and base numbers consisting in oil titration with potassium hydroxide is unsuitable in producing built-in sensors.

The diagnosis device built into the lubrication system is based on the determination of the acid number by electrochemical method [10, 11]. The device consists of two steel electrodes coated with inert metals, e.g. gold or platinum (Fig. 1.4). The oil under testing fills the gap between the electrodes which has low electric resistance because its thickness is small (150 μm). The electrodes form a capacitor to which an alternative voltage of triangle shape is applied from an external source.

Fig. 1.4 Diagnosis device based on the assessment of the acid number of oil

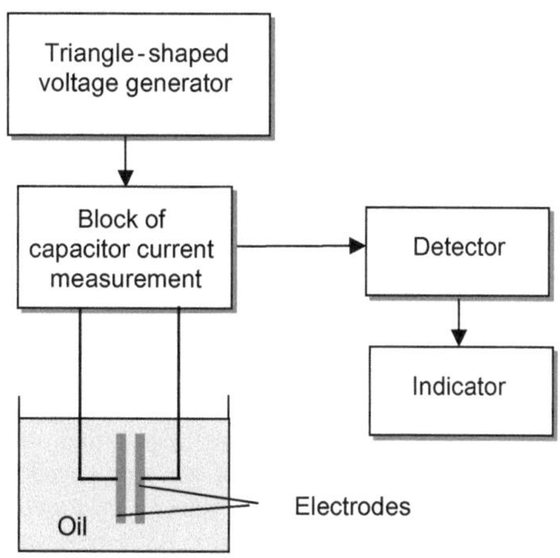

The current variation at the output from the capacitor is governed by the oil electrochemical activity (its acidity), which causes electrochemical reactions on the electrode-oil interface.

The method does not allow separating the acids formed in oxidation from those that are found in the additives. Acidity of the additives governs a high acid number and with the decomposition of the additives the oil acidity decreases thus distorting the validity of the oil quality evaluation.

Alkali number sensors. Alkali additives are introduced into the motor oil for decreasing its acidity. With time the additive content decreases and the remaining alkali additives as well as all the base components in the oil are characterized by the total base number TBN. It is evaluated in laboratory with the potassium hydroxide required to neutralize the base components and this number is efficient to determine the oil aging.

The principle of the sensor for prompt diagnosis of oil operation capacity [12] is determined by the interrelation between TBN and sound velocity in oil (Fig. 1.5a). As is shown in the figure the sound wave velocity increases with decreasing base number. Using this relation with taking account of the oil temperature the sensor makes it possible to evaluate TBN by the ultrasonic wave velocity. The sensor comprises the transmitter-receiver of ultrasonic-transducer *3* (Fig. 1.5b), which is mounted in case *5* and radiates ultrasound waves normally to the reflecting surface of transducer *2* and oil surface *1*. The reflected waves are transformed by element *3* into electric signals carrying information on the time of ultrasonic wave propagation reflected from surfaces *1* and *2*. Then the signals are fed to block *6* of the measurement of wave propagation time where the oil level and wave velocity are evaluated. Then the signal proportional to the wave velocity is fed to block *10* of temperature correction where the data on the oil temperature are fed from temperature sensor *4*. After temperature correction the signal goes to block *11* for determination of the oil quality.

Sensor has a disadvantage, because it is provided with a complicated temperature correction, so it should be equipped with a highly accurate thermometer with a wide range of the operation temperatures.

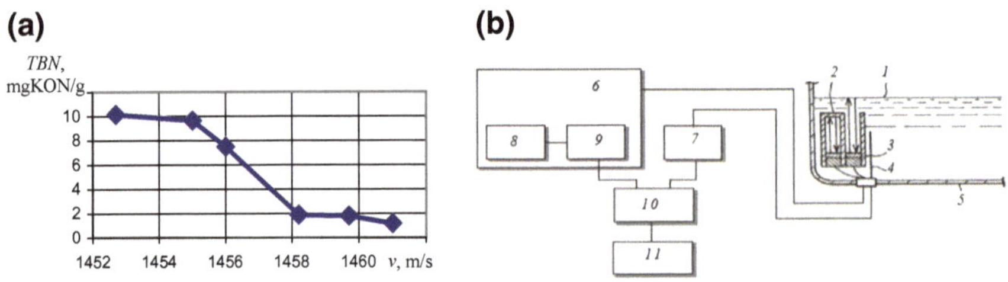

Fig. 1.5 Interrelation between the acid number of oil and the sound velocity in it (**a**): *1* oil surface, 2 reflecting surface, *3* receiving-transmitting element of the transducer, *4* temperature sensor, *5* case, *6* measurement block of wave propagation time, *7* time measurement block, *8* oil level measurement block, *9* sound velocity measurement block, *10* temperature correction block, *11* block of oil quality determination; ultrasonic sensor schematic (**b**), [12]

The sensors based on the *piezo-chemical-electrical method* were also developed basing on the experience of using the acoustic sensors of gas and liquid [13]. They use piezoresonator (Fig. 1.6) consisting of quartz crystal *3* where surface electrodes *2* are applied. Special layer *1* (polymer or metal) that is chemically active to the test medium component (alkali or acid) is applied to one of the surfaces comprising the electrode. Selective sensibility of the layer allows concentrating ("binding") molecules of the test substance, which causes variation in mass and viscoelastic properties. The variation results in shifting resonance frequency of the quartz resonator. The processor compares the resonance frequency to the basic one and determines the mass of the test component by their difference.

The oil quality sensor [14] based on this principle allows determining one or several components in oil. The sensitive layer (about 1.5 μm thick) is made of polymers, in particular, polyurethane. The products of oil decomposition are built into the polymer structure. This can be realized, e.g. by using alkali monomers or fluoropolymers with alkali residues. At immersing the sensor into the oil the molecules of the test substance diffuse into the sensitive layer. To compensate the effect of the oil viscosity variation the sensor is equipped with the basic quartz resonator without the sensitive layer.

The disadvantage of the sensor that impedes its usage is a necessity to replace the sensible layer after it is contaminated.

Sensors based on the monitoring of optical properties. The advantage of optical sensors over the electric ones is that they are efficient in aggressive media and are not subjected to the effect of electric and magnetic fields.

Sensors based on absorption/transmission of optical radiation. Molecular spectroscopy, particularly Fourier Transform Infrared FTIR-spectroscopy, recently has been widely used to evaluate a number of quality parameters of oil, viz. content of water, fuel, coolant liquid (glycol), carbon black, oxidation products and additives.

The principle of molecular spectroscopy is based on the fact that individual molecules absorb optical radiation at some characteristic resonance frequencies. These frequencies result from the presence of molecular groups of two or more atoms. Thus, in a water molecule O–H is such a functional group. It contributes to the resonance absorption, in particular, at a frequency of 3450 cm^{-1}.

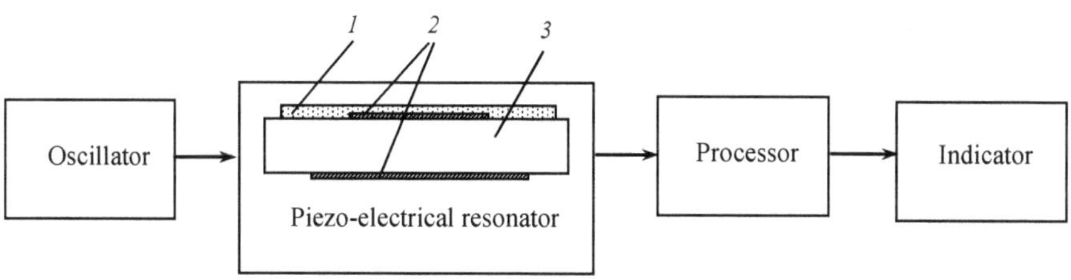

Fig. 1.6 Piezo-electrical sensor of oil quality [13]: *1* polymer functional layer; *2* electrodes; *3* quartz crystal

Laboratory FTIR spectrometers make possible analysis of the oil within 600–4000 cm^{-1} frequencies [15].

FTIR portable sensors based on a small-size IR-cell have been developed that are used for oil analysis in real time. The current FTIR methodology is based on ASTM Standard Practice E2412-10 and is restricted primarily to petroleum or mineral oils [16]. It is used in gas turbines and diesel engines to monitor basic characteristics of oil (acid and base numbers, thermooxidative aging, antioxidative and antiwear additives).

Regardless of a large scope of information obtained by IR-spectrometry of oil it is not always valid. It results from the fact that some molecules have similar functional groups (e.g. molecules of three substances contained normally in the worked oil, viz. water, glycol, and phenol antioxdative additive). Since most worked oils are a complex mix of different molecules including those of the oil basestock, additives, products of oil aging, wear and contamination particles, then the IR-spectrum of oil can be interpreted with some uncertainty. Moreover, a wide usage of IR-spectrometers is impeded by their high cost.

Therefore, for prompt diagnosis simpler and cheaper devices are developed that use near IR-radiation. Thus, a method and device for diagnosis of the oil aging of automobile engines is available [17, 18]. It is based on the measurement of optical density of oil on two wavelengths of near IR-range. Characteristic spectral dependencies of IR-radiation losses per unit length of the optical path in oil for different degrees of the oil aging are shown in Fig. 1.7a. The losses in the visible region increase drastically so much that even at a small length of the optical path in oil the intensity of optical radiation becomes lower than the sensitivity threshold of the photoreceiver. However, a high cost and large overall dimensions of the radiation sources impede the use of mean and long IR-radiation. Therefore, a short IR-range is used in the sensor. The difference in losses for two wavelengths characterizes the slope of *A–A′* line at the initial stage of oil quality deterioration, that of *B–B′* line at the mean stage, and *C–C′* at the last stage of the oil aging. Therefore, the difference in the losses for the two wavelengths can be used to assess the oil condition. The device comprises IR-source *1* with two wavelengths λ_1 and λ_2 (Fig. 1.7b). By optical waves *4*, light guide *6* installed into sensor *5*, and mirrors *7* the radiation passes through gap *8* 0.5…–2.0 mm wide filled with oil under analysis *9*. Then optical radiation by the photoreceiver *2* is transformed into electrical signal, which goes to the processor. Processor *3* computes absorption α of IR-radiation in oil for the beams of two wavelengths λ_1 and λ_2 and determines the difference in losses $\Delta_\alpha = \alpha_{\lambda 1} - \alpha_{\lambda 2}$. The degree of the oil aging is assessed proceeding from the calibration dependence between the aging degree and difference Δ_α.

The technique is used in the oil condition sensor produced by Windrock Co.

The IR-detector disadvantage results from thin measuring gaps with the test oil, which causes a rapid contamination of the optical windows. In addition, receivers operating in a spectral region >1000 nm are rather expensive.

Fluorescent technique. In addition to IR-spectroscopy fluorescent spectroscopy is used in oil condition monitoring. Fluorescence is a light radiation of molecules in their transition from electronic excited state to the ground one. In this case the

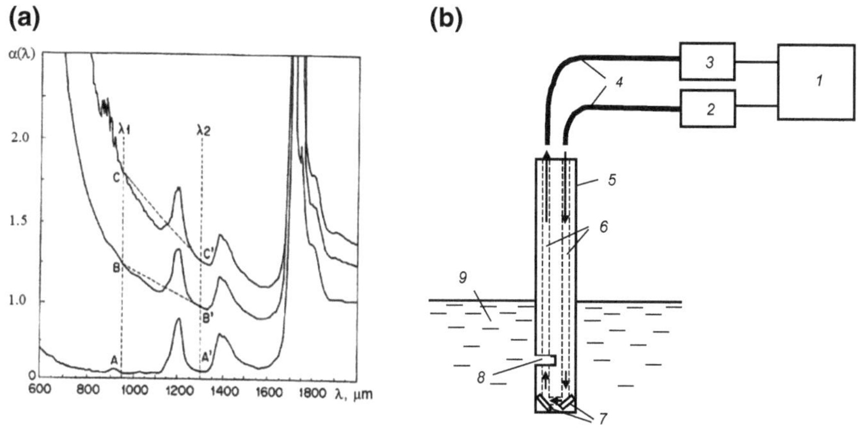

Fig. 1.7 Change in IR-radiation absorption spectrum in oil aging **a** and diagnosis device based on measuring optical density of oil **b**: *1* processor; *2, 3* source and receiver of IR-radiation; *4* optical fibers; *5* sensor; *6* light guide; *7* mirrors; *8* measuring gap; *9* oil under testing, [17]

radiation spectrum normally lies in the range from UV-to visible wavelengths and sometimes in the near IR-range. The molecules of fluorophores being in the ground state when excited with external radiation (of visible or UV-range) absorb the photon and transfer at some higher oscillating energy levels. In a short period of time the flurophore molecule returns to its basic state radiating the photon. The energy of the photon radiated is always less than that of the photon absorbed because of energy dissipation. Also fluorescence occurs at a longer wave than absorption. The shift of the wavelengths is about 50–70 nm. In a general case fluorophore concentration C can be assessed by fluorescence power P from the relationship $P = QP_0(1 - t^{-\beta Cl})$ where Q is quantum yield (ratio of the number of radiated photons of fluorescence to the number of absorbed photons), P_0 is the power of incident power, β is molar absorptivity of the fluoroscenting component, l is the length of the optical path of radiation in the substance under testing.

The fluorescence spectra of the substance are analyzed by fluorescent spectrometers.

Typical fluorescent spectrometer (Spex, Fluoromax) consists of two monochromators scanning the exciting and emitted radiation [19]. The source of exciting radiation is a 150 W-xenon lamp, which provides excitation within a 200–900 nm range. The sample under testing is lighted with radiation transmitted from the source along the optical fiber with a wavelength chosen by a monochromator. Fluorescence radiation is transmitted from the source along the fiber to the input slit of the second monochromator and transformed into the electrical signal by a photomultiplier. Then the electrical signal is computer-processed.

Possible application of fluorescence spectroscopy to monitor the oil condition based on polyphenyl ethers with different antioxidative additives was discussed elsewhere [20]. The oil oxidation was shown to cause the variation in its fluorescence characteristics, which can be used in diagnosis of the oil efficiency.

One should note that fluorescent monitoring is not applicable to all mechanisms. In particular, fluorescent analysis is unsuitable in automobile engines. It results from the fact that fuel contaminating the oil in operation contains a large quantity of aromatic hydrocarbons than oil, which affects considerably the oil fluorescence [21].

Even though fluorescent spectroscopy does not allow assessing the concentration of individual components it is informative as far as the general "integrated" oil condition is concerned. In those cases when information is sufficient to assess the oil efficiency the methods of fluorescent analysis are helpful in producing portable sensors built into the lubrication system. In such sensors cheap quartz (and even polymer) optical fibers and UV-diodes can be used.

1.4 Fluorescence Methods and Tools for Real-Time Oil Oxidation Monitoring

Fluorescence of oil results mainly from the benzene rings and compounds with a large number of mated bonds. The intensity and spectrum of radiation of oils in their aging vary. Figure 1.8 presents the monitoring results of synthetic aviation oil Hatco Corporation (MIL-L-23699D) [19]. The oil has experienced artificial aging at the laboratory conditions by heating it up to 160 °C with permanent mixing. To simulate the high-humidity medium the water-saturated air passed through the oil with rate 10 cm^3/min. The oil aged for 60 days and the samples were sampled for analysis each 1–4 days. The fluorescence spectra were obtained for each oil sample on the Spex (Fluoromax) spectrometer at an exciting wavelength of 360 nm. It is seen from the figure that the fluorescence spectrum varies in oil degradation.

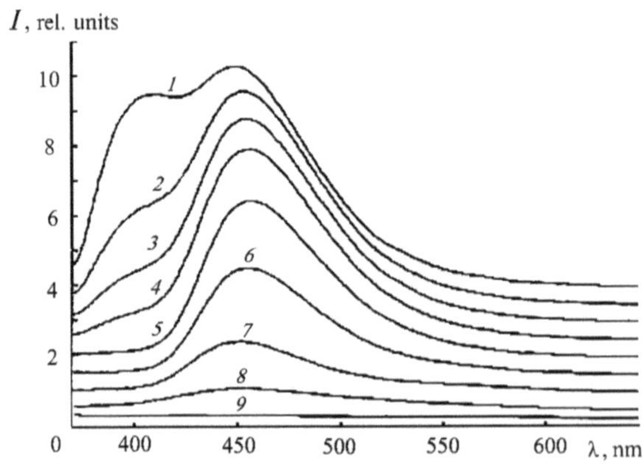

Fig. 1.8 Change in fluorescence spectra of synthetic aviation oil Hato Corporation (MIL-L-23699D) as a function of oxidation time: *1* 1; *2* 2; *3* 3; *4* 4; *5* 5; *6* 11; *7* 18; *8* 52; *9* 60 days, [19]

The main techniques applied in fluorescence equipments are measurement of fluorescence intensity and measurement of fluorescence lifetime.

Fluorescence intensity technique. The increase or decrease in fluorescence intensity is measured and then correlated to the analyte concentration, that is, the fluorescence intensity is proportional to the amount of analyte.

Change in fluorescence intensity is often caused by quenching. Degree of quenching is expressed as a function of analyte (quencher) concentration using the Stern-Volmer equation:

$$\frac{I_0}{I} = 1 + k(C_q),$$

where

- I_0, I intensity of fluorescence without a quencher and with a quencher, respectively,
- k is the Stern-Volmer quenching constant and
- C_q is the concentration of the quenching species.

Fluorescence lifetime technique. The fluorescence lifetime is fundamentally different parameter than the more familiar fluorescence intensity. Formally, the lifetime is the average amount of fluorescence time spent in the excited state between absorption of a photon and its emission.

There are two main approaches to measure the lifetime: time and frequency domain.

The time domain measurement embodies determining the time-dependence of the fluorescence emission following a brief flash of excitation. The light intensity emitted from a molecule exited by a short pulse of light decays exponentially with time. The decay time pattern is unique for each molecule and can be used for analytical purposes.

Frequency domain instruments (phase fluorometers) provide rapid readout of two lifetime-related parameters, the phase and modulation. In frequency-domain fluorometer, the sample is excited with light whose amplitude is modulated sinusoidal at a frequency close to the reciprocal of the lifetime, typically about 1–300 MHz. The fluorescence emission is also modulated at the same frequency, but it is phase-delayed. The phase shift is expressed as a phase angle from which the lifetime can be determined using the simple relationships between the modulation frequency and the degree of demodulation. The concentration of analyte that induces change in the molecule's fluorescence lifetime can be determined by measuring phase angle values [20].

Various modifications of these main techniques are developed, some of which could be employed for oil condition test.

The synchronous scan fluorescence spectroscopy technique produces spectra resulting from scanning both the excitation wavelength and the detection wavelength with a fixed wavelength separation. By using this technique [21] it is possible to distinguish oils belonging to the same class from each other, but the distinguishing ability is still not adequate enough to discriminate oils of closer grades. In addition, this technique cannot be practically used in remote sensing since it is not easy to tune a high-intensity laser over a wide range of excitation wavelengths.

The contour (total luminescence) spectroscopy technique [22] is another technique that can be used for the purpose of crude oil characterization. It produces contour diagrams of oils that are constructed out of many emission spectra each of which is excited at a different wavelength.

The laser-induced fluorescence technique that promises a good distinguishing ability and, at the same time, a practical remote sensing application is the time-resolved laser-induced fluorescence technique. The suggestion of this technique as a tool for oil characterization was made as early as 1971 by Fantasia et al. [23], who recommend the use of lifetime measurements as an additional tool for crude oil characterization.

The variation of the fluorescence decay time as a function of wavelength across the emission profile for a variety of materials could be used to discriminate between very similar substances, i.e., it could be used as a tool for correct characterization.

For example, the authors of US6633043 [24] propose a method based on time-resolved, laser-induced fluorescence spectroscopy for characterization of oils without taking into account their relative intensities. Such technique can be useful in monitoring of the degradation of lubricant and transformer oils basing on the fact that shapes of the time-resolved fluorescence spectra of the fresh and the degraded oils vary in different manners.

A method for monitoring the degradation of lubricants and transformer oils using the normalized time-resolved fluorescence comprising the following steps:

- exposing an unknown lubricant or transformer oil sample to a pulse of ultraviolet laser radiation;
- measuring the intensities of resulting fluorescence over the spectrum of wavelengths of light from oil sample at specific narrow time gates within the temporal response of the laser pulse to form a time-resolved spectrum;
- normalizing the time-resolved spectrum at a particular emission wavelength;
- plotting the time-resolved spectrum in contours as functions of wavelength and time simultaneously;
- comparing the resultant plots of the plotting step with those of similar samples taken at known levels of degradation; and
- characterizing the unknown sample based on similarity of the resultant plots with those of a known sample.

A car-engine lubricant (Fuchs Super GT 20W/50) was chosen to investigate the degradation time-resolved fluorescence properties. The samples were fresh, 5-week, and 4-month degraded samples, and their time-resolved fluorescence spectra were excited and measured using the experimental setup.

Measuring system (Fig. 1.9) includes pulsed UV laser irradiating via mirror the oil sample contained in a quartz cuvette. The resulting fluorescence signal from the

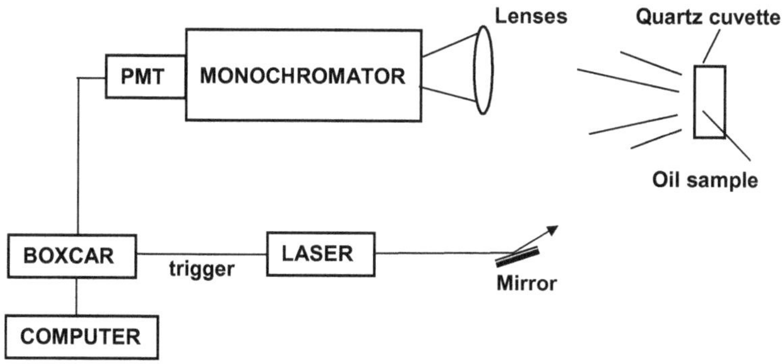

Fig. 1.9 Experimental setup for measurement of time-resolved fluorescence spectra, [24]

oil is steered by quartz collecting lenses onto the entrance slits of a medium-resolution monochromator for dispersion, after which it is detected by a fast photo multiplier (PMT) mounted at the exit slits of the monochromator. The detected fluorescence signal is sent to the signal processor coupled with the gated integrator (Boxcar) and a PC for sampling and digitizing it according to specific time gate widths.

Figure 1.10 show normalized time-resolved fluorescence spectra measured at Time Gate TG 0 and TG 10 of the lubricant oil when it was fresh, 5-week, and 4-month degraded. The spectra were normalized at 450 nm and excitation wavelength was equal 280 nm. It is clear that the spectra at these two time gates vary in shape in different manner depending on the ageing state of the oil. By considering

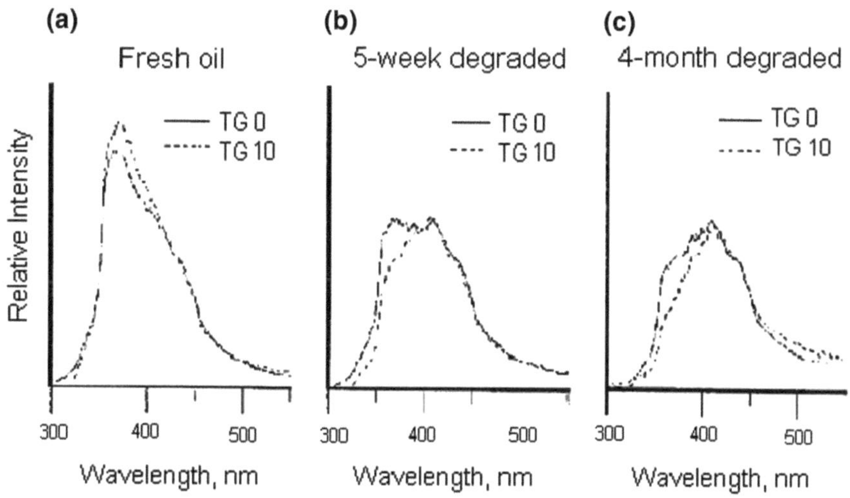

Fig. 1.10 Normalized time-resolved fluorescence spectra at Time Gate TG 0 and TG 10 for fresh **a**, 5-week degraded **b**, and 4-month degraded (**c**) lubricant oil, [24]

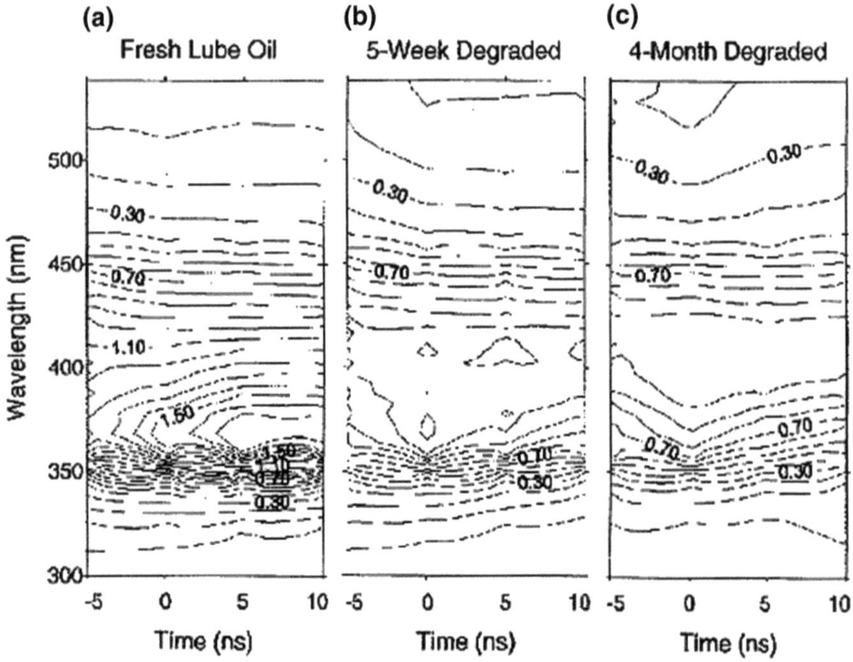

Fig. 1.11 Contours of equal fluorescence intensities of the normalized time-resolved fluorescence spectra of fresh **a**, 5-week degraded **b**, and 4-month degraded (**c**) lubricant oil, [24]

the normalized spectra of the other time gates and comparing certain areas under their curves it is possible to predict the ageing of the oil quantitatively.

It is also possible to predict the ageing of the oils using the contour diagrams constructed upon normalized time-resolved fluorescence spectra. Figure 1.11 present contours of equal fluorescence intensities of the normalized time-resolved fluorescence spectra of the oil in the three ageing cases. The contour lines reflect the variations in the shapes of the time-resolved fluorescence spectra, but not their relative intensities. Normalization wavelength was 450 nm.

A comparison between these diagrams shows that there is general trend in the contour patterns near 350 nm and also in the region between 350 and 430 nm, which change gradually with respect to the ageing of the oil. As the oil ages the contour lines between 330 and 370 nm become less congested, and the contour lines of higher intensity level, e.g., 1.10 and 1.50, start to disappear in the area between 350 and 430 nm. It is possible to use these diagrams in predicting when the aged oil needs to be replaced by fresh oil.

So, we could see that the main techniques applied in fluorescence equipment have limitations. Disadvantage of technique of fluorescence intensity measurement is dependence of the intensity on oil absorption because oil has rather high optical density. In this case, a special measurement technique should be used to eliminate this influence. Techniques (particularly, time-resolved fluorescence spectroscopy) based on the fluorescent molecule exited state lifetime are free from this disadvantage but is rather complex for on-line monitoring. Therefore, development of new fluorescent techniques, which is simple for realization in on-board application, is a topical question.

1.5 Fluorescence Emission Ratio Technique and Sensor for On-Board Application

The fluorescence of oil as any organic substance is caused by energy transitions in π-orbitals of C=C bonds. In addition, coupled π-systems require lower excitation energies than isolated bonds due to a higher mobility of π-electron and hence they are excited more easily. Therefore, aromatic compounds are the main sources of the fluorescence of organic substances [25].

The growth of the number of cycles of aromatic compounds or the length of double bond chains increases the mobility of π-electron and shifts the absorption and emission spectra towards the longer wavelengths. Molecular interactions affect considerably the fluorescence intensity. Intramolecular energy transfer can decrease or even annihilate the fluorescence of a molecule, while intermolecular interactions govern the fluorescence quenching.

Oil oxidation causes the rapid saturation of unsaturated aromatic molecules with oxygen-containing groups. As a rule, the oxidation of aromatic molecules leads to the formation of oxygen bridges (alkyl and aryl ethers) and general increase in polarity. All these factors result in higher delocalization of π-electron that increases the relative number of emitting centers in the long-wave region and for this reason the emission spectrum shifts towards the longer wavelengths.

This phenomenon of the fluorescence spectrum shift caused by oil oxidation has formed the ground for the method of the on-line monitoring of oil condition [26].

Figure 1.12 shows the changes in the spectra of the relative fluorescence intensity measured with the fluorometer Multifrequency Cross-Correlation Phase and Modulation (Model K2, ISS Co., USA) at the excitation wavelength of 300 nm for fresh hydraulic oil DTE-24 $\left(I_{fr}^{*}(\lambda)\right)$ and two samples (A and B) of used oil $\left(I_{A}^{*}(\lambda), I_{B}^{*}(\lambda)\right)$, the operating time of oil B being longer than that of oil A. The relative fluorescence intensity decreases in the short-wave region and increases in the long-wave one, i.e. with prolonging the duration of oil operation the spectrum shifts towards the longer wavelengths.

To evaluate the oil oxidation degree we proposed the diagnostic parameter F (fluorescence emission ratio) that characterizes the shift of the spectrum as the ratio of the fluorescence intensity $I_{\Delta\lambda l}$ in the longer wave range to the fluorescence intensity $I_{\Delta\lambda sh}$ in the shorter wave range [26, 27]:

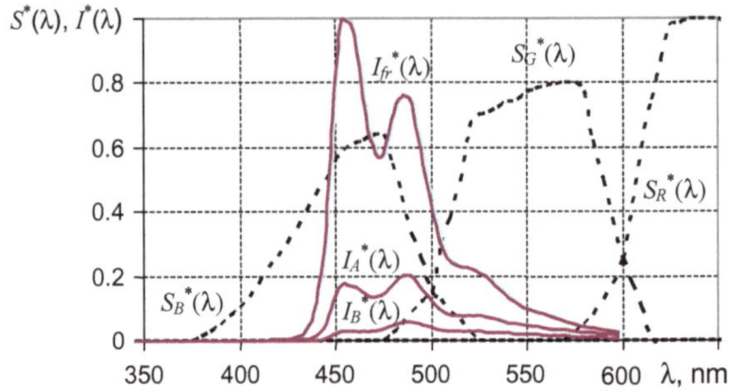

Fig. 1.12 Spectra of relative fluorescence intensity of fresh oil $\left(I_{fr}^{*}(\lambda)\right)$ and two samples (*A* and *B*) of worked $\left(I_{A}^{*}(\lambda), I_{B}^{*}(\lambda)\right)$ hydraulic oil DTE-24 at excitation wavelength 300 nm and relative spectral sensitivity of RGB photodiode in *red* $S_{R}^{*}(\lambda)$, green $S_{G}^{*}(\lambda)$, and *blue* $S_{B}^{*}(\lambda)$ ranges

$$F = \frac{I_{\Delta\lambda l}}{I_{\Delta\lambda sh}}. \tag{1.1}$$

The shift of the fluorescence spectrum towards the long-wave range depends on the oil oxidation degree and does not depend on the oil temperature and optical density.

The fluorescence intensity in the short- and long-wave ranges is measured with the color sensor (RGB photodiode). It is capable of measuring simultaneously the intensity of oil fluorescence in three spectral ranges, i.e. red, green, and blue. For example, one of such detectors is Color Sensor MCS3AT (MAZeT GmbH, Germany) whose relative spectral sensitivities in the red $S_{R}^{*}(\lambda)$, green $S_{G}^{*}(\lambda)$, and blue $S_{B}^{*}(\lambda)$ ranges are shown in Fig. 1.12.

The efficient spectral fluorescence intensity $I_{eff}(\lambda)$ (the spectral intensity transformed into the electric signal by the RGB photodiode) is the product of the spectral fluorescence intensity $I(\lambda)$ of oil and the spectral sensitivity $S(\lambda)$ of the photodiode:

$$I_{eff}(\lambda) = I(\lambda)S(\lambda). \tag{1.2}$$

Figure 1.13 presents the calculated relative efficient spectral intensity of fluorescence $I_{eff}^{*}(\lambda) = I_{eff}(\lambda)/I_{\max}$. The efficient spectral intensity in the red range ($\Delta\lambda$ = 590–750 nm) is zero and the ratio of the efficient fluorescence intensities in the green ($\Delta\lambda$ = 490–610 nm) and blue ($\Delta\lambda$ = 400–510 nm) ranges increases as the duration of oil operation becomes longer.

The output signals of the RGB photodiode in the red J_R, green J_G, and blue J_B ranges are proportional to the efficient fluorescence intensity (the area under the curves of the efficient spectral fluorescence intensity) and are found from the formulas:

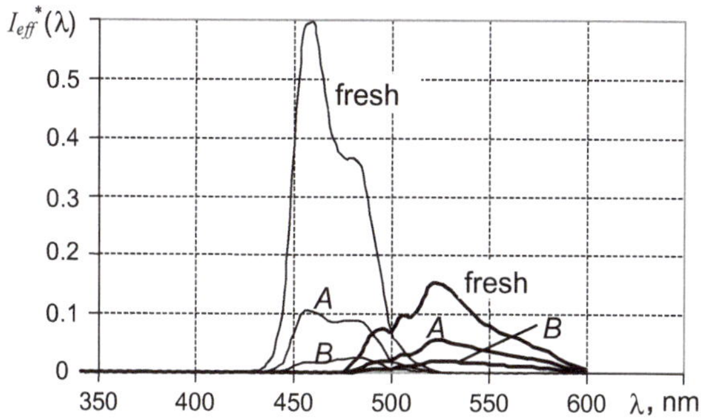

Fig. 1.13 Relative efficient spectral fluorescence intensity of fresh oil and two samples (*A* and *B*) of worked hydraulic oil DTE-24

$$J_R = \int_{\Delta\lambda_R} I_{eff}(\lambda)d\lambda = \int_{\Delta\lambda_R} I(\lambda)S(\lambda)d\lambda = I_{\max}S_{\max} \int_{\Delta\lambda_R} I^*(\lambda)S^*(\lambda)d\lambda, \quad (1.3)$$

$$J_G = \int_{\Delta\lambda_G} I_{eff}(\lambda)d\lambda = \int_{\Delta\lambda_G} I(\lambda)S(\lambda)d\lambda = I_{\max}S_{\max} \int_{\Delta\lambda_G} I^*(\lambda)S^*(\lambda)d\lambda, \quad (1.4)$$

$$J_B = \int_{\Delta\lambda_B} I_{eff}(\lambda)d\lambda = \int_{\Delta\lambda_B} I(\lambda)S(\lambda)d\lambda = I_{\max}S_{\max} \int_{\Delta\lambda_B} I^*(\lambda)S^*(\lambda)d\lambda. \quad (1.5)$$

The calculated output signals of the RGB photodiode are shown in relative units in Fig. 1.14a.

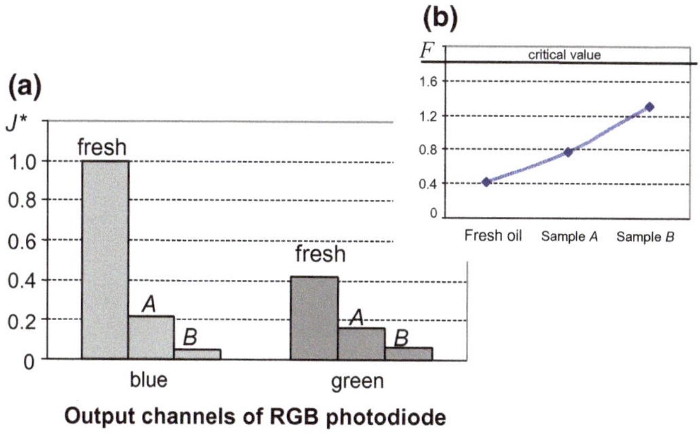

Fig. 1.14 Output signals of RGB photodiode in relative units **a** and fluorescence emission ratio parameter *F* (fluorescence emission ratio) as function of oil operation duration **b**

In order to calculate the diagnostic parameter F it is required to determine preliminarily the ranges $\Delta\lambda_l$ and $\Delta\lambda_{sh}$ when analyzing a specific type of pure oil by the following criterion. If the spectrum of pure oil occupies mainly the green and red regions, i.e. $J_R > J_B$ and $J_G > J_B$ then $\Delta\lambda_{sh} = \Delta\lambda_G$ and $\Delta\lambda_l = \Delta\lambda_R$. If the signal is less intensive in the red range than in the blue and green ones then $\Delta\lambda_{sh} = \Delta\lambda_B$ and $\Delta\lambda_l = \Delta\lambda_G$. The latter condition is fulfilled in the example under consideration and, consequently, the parameter F is

$$F = \frac{I_{\Delta\lambda l}}{I_{\Delta\lambda sh}} = \frac{J_G}{J_B}. \tag{1.6}$$

Variation in the parameter F with increasing the oil operation duration is shown in Fig. 1.14b. According to the monitoring technique the parameter F is compared to its preset critical value for a particular mechanism and finally the conclusion is drawn on the oxidation degree and efficiency of oil.

A simple detector Fluor–2 based on this phenomenon, as shown in Figs. 1.15 and 1.16 was developed in this work with a three-color sensor as photo-receiver that detects optical light intensity in three wavelengths ranges–red, green, and blue (RGB sensor).

The radiation from the UV diode (1) passes through the bifurcated optical fiber and optical window (9) to the test oil. A ball lens (5) is used to focus the UV diode radiation at the optical fiber end. The optical fibers are mounted in the fiber holder (13), which is inserted into the housing (12). An optical window is fixed to the housing (12) by a nut (8). The optical fiber transmits the fluorescent emission from the test oil to the RGB sensor (14). A photodiode (2) is applied to stabilize the UV radiation feedback. The UV diode, RGB sensor, and feedback diode are cable-connected to an electronic block (not shown in Fig. 1.16). O-rings (7) and (11) prevent oil leakage into the sensor. A cover (4) is used to protect the detector. The electronic microcontroller, amplifier and power PCBs, and an LCD panel are housed in a box, cable-connected to the detector probe (15). The fluorescence emission from the test oil is measured on a RGB sensor (14), and then the parameter F is calculated by an embedded microprocessor in the device. Value $I_{\Delta\lambda l}$ was selected as I_G and $I_{\Delta\lambda sh}$ was selected as I_B, for the evaluation of the F [27].

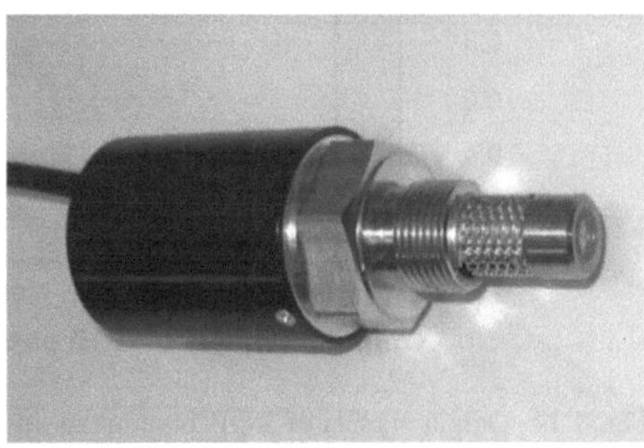

Fig. 1.15 View of fluorescent detector Fluor-2

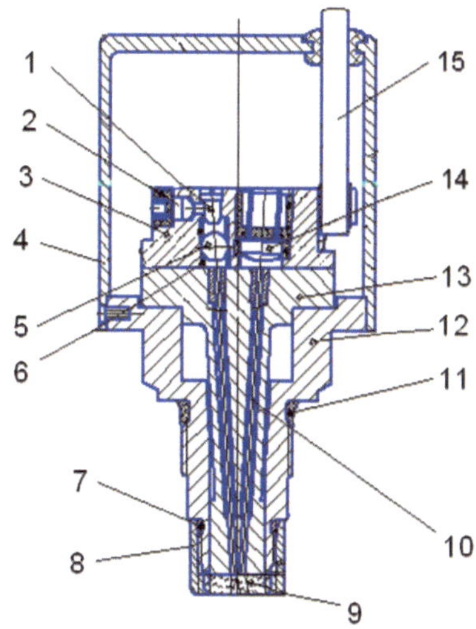

Fig. 1.16 Fluor–2 detector design with optical fibers: *1* UV diode, *2* feedback photodiode, *3* bush, *4* cover, *5* ball lens, *6* insert, *7* O-ring, *8* nut, *9* optical window, *10* optical fibers, *11* O-ring, *12* housing, *13* fiber holder, *14* RGB sensor, *15* electric cable

1.5.1 Application of the Fluorescent Sensor for Hydraulic Oil Condition Monitoring

The fluorescent detector Fluor-2 was tested using the UVdiode LED3-UV-395-30 (Bivar Inc., USA) with the wavelength of 395 nm as the exciting emission source and the RGB photodiode Color Sensor MCS3AT as the detector. The sensor can be used as a portable device under laboratory conditions or built in the oil line. The data are recorded and processed according to software driven by a microcontroller. When the diagnostic parameter *F* exceeds its critical value the signal (light diode of the electron unit) flashes, which means the oil should be replaced.

The method efficiency was evaluated in laboratory study when testing pure and worked mineral hydraulic oils, i.e. Teresstic T-100 (ESSO) taken from a hot rolling mill, oil Rando HD-32 (TEXACO) taken from the input unit of a hydraulic system, and oil DTE-24 (MOBIL) and synthetic oil Cosmolubric HF-130 (Houghton) taken from a hot rolling mill. Measured values of the parameter *F* were compared to the total acid number found by the standard laboratory method (ISO 6618).

The measurement results (Fig. 1.17) demonstrate a good correlation between the diagnostic parameter *F* and the total acid number TAN that indicates a high validity of the results of estimating the oxidation degree of hydraulic oil by the fluorescent method.

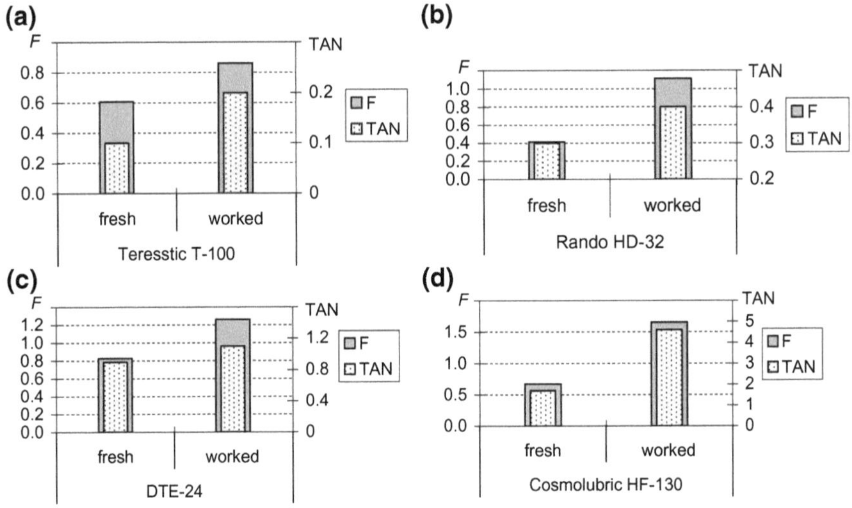

Fig. 1.17 Correlation between diagnostic parameter *F* and total acid number TAN when analyzing oils: **a** Teresstic T-100; **b** Rando HD-32; **c** DTE-24; **d** synthetic oil Cosmolubric HF-130

1.5.2 Application of Fluorescence Emission Ratio Technique for Transformer Oil Monitoring

Similar to industrial oils, insulating transformer oils are oxidized under the influence of excessive temperature and oxygen, particularly in the presence of small metal particles that act as catalysts and which result in an increase in the acid number owing to the formation of carboxylic acids [28]. Further reactions can result in sludge and varnish deposits, and additionally, in increased acidity has a damaging effect on the insulating paper. It is known that transformer oil degradation also produces charged by-products, such as acids and hydro peroxides, which tend to reduce the insulating properties of the oil. An increase in the acid number is often accompanied by a decrease in dielectric strength and increased water content. For these reasons transformer oil needs condition monitoring which can be based on the same principles as industrial oils.

Methods of analyzing physical and chemical parameters of insulating oil, such as color, water content, and the acid number, have been used to determine oil degradation. These methods have formed a core part of the predictive maintenance of electrical insulating oil [29], but they are limited to analysis in laboratory, whereby it is not possible to provide an early warning before any failure, so on-line monitoring technology for evaluating insulating oil degradation is crucial.

Fluorescence spectroscopy is one of techniques which can satisfy the requirements of off-line and on-line transformer oil monitoring [30].

Transformer oils are mainly refined from petroleum, but they are very complex mixtures and may consist of as many as 2900 paraffinic, naphthenic, and aromatic (25%) hydrocarbon molecules while the lattes ones are fluorescent [31].

An apparatus was developed to measure the oxidation of electrical insulating oil and it can detect the intensity of fluorescence light reflected from the oil in real time, and this intensity is an indication of oil oxidation based on the UV fluorescence emission ratio. The test results were compared with other measurement methods such as the titration method and IR spectroscopy.

Oxidation test results. The test oil was Mictrans KS 2301, 1-2 electrical insulating oil (Michang oil, kinematic viscosity: 9.00 cSt @ 40 °C, initial TAN value: 0.0055 mgKOH/g, water content: less than 15 ppm, life criteria: TAN less than 0.24) [32].

For preparing an artificially oxidized oil sample it was oxidized by ASTM 2440-99, with some modifications, as follows: 500 ml of test oil was heated to 110 ± 0.5 °C, and mixed with 1.5 l/h of 99.4% pure oxygen in the customized oil apparatus, as shown in Fig. 1.18. After the oxygen rushed into the drying tower, it entered the micro-bubbler and contacted the copper coil (180 cm long). The oil was stirred at 300 rpm for uniform oxidation.

Every 12 h, a small portion of the test oil sample was extracted from the apparatus in order to measure the level of oxidation of the oil. The oxidation levels of test oil samples were also measured using two types of commercially available lubricant monitoring sensors (TAN test cell of Kittiwake Co. and Fluidscan of Spectro Co.). Fresh oil and oil after 822 h of aging were also analyzed by FTIR of Perkin Elmer Co., UV-VIS spectrophotometer (Opron-3000, Hanson Technology) and dielectric dissipation factor (Tan(delta)) was measured by SOKEN DAC-5016A (test condition: 60 Hz, 80 °C; test voltage: 250 V; test method: KS C

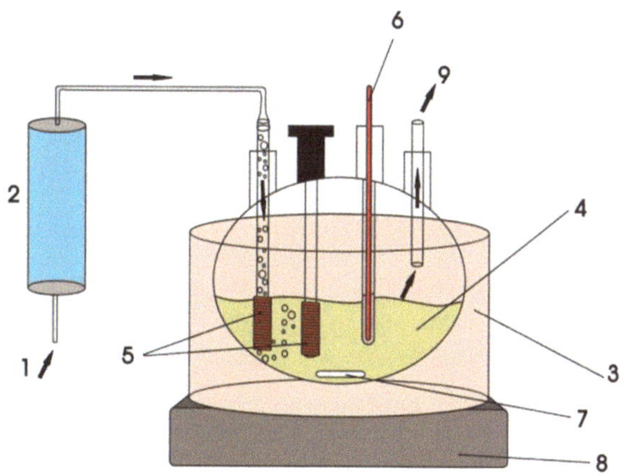

Fig. 1.18 Test oil oxidation apparatus: *1* oxygen in, *2* drying tower, *3* aging bath, *4* test oil, *5* copper catalyst, *6* thermocouple, *7* magnetic stirrer, *8* heater and stirrer, *9* oxygen out

2101). Simultaneously optical densities (D_R, D_G and D_B) in R, G and B wavebands were measured by the RGB sensor of Fluor–2. The TAN test cell (FG-K25196-KW) uses a titration method based on ASTM D974-11. It operates by measuring the color change caused by acid in the test oil sample. In this work, a 20 μL pipette was used for TAN test in order to increase the titration accuracy. Fluidscan®Q1000 is a handheld mid-infrared spectrometer based on ASTM standard practices (E2412-04), that measures the total acid number (TAN), oxidation level, and other related parameters. After the measurement, the oil sample was returned to the bath with test oil (4).

Test results were summarized in Table 1.3 and Figs. 1.19, 1.20, 1.21 and 1.22. It was found that parameter TAN and parameters of Fluidscan show the oil degradation started in 550 h of oil oxidation test. Once oil oxidation was initiated, the acid number increased abruptly, as shown in Figs. 1.19 and 1.20. FTIR analysis of oil sludge samples formed at 822 h clearly showed a significant increase in the carbonyl peaks (–C=O, near 1700–1750 cm^{-1}) and hydroxide peaks (–OH, near 3300–3500 cm^{-1}), which are indications that the oil was oxidized, as shown in Fig. 1.21. Decrease of insulating property of the oxidized oil at 822 h is confirmed by the increase in dielectric loss tangent (from 0.001 to 2.55%). It increases with the introduction of polar molecules such as esters, aldehydes, and ketones. As it is evident from UV–VIS spectra (Fig. 1.22), a prominent absorbance waveband 370–430 nm appeared in the spectrum of oxidized oil sample during 822 h. From the Table 1.3, we can see that optical densities Db of fresh and oxidized oil measured in the range 400–510 nm are in close correlation with UV–VIS spectroscopy data. It is obvious that the absorption in this waveband can be used to monitor transformer oil condition particularly by on-line measurement of oil optical density.

It was found that the *F* parameters measured by Fluor–2 showed very similar results with those obtained from Fluidscan®Q1000 and titration methods. The *F* parameter seems to increase relatively earlier (in 450 h) than others, providing an early warning of oil oxidation, since *F* parameter is more integrated reflecting the

Table 1.3 Measurement results

Oxidation Time, hours	ΔF	Fluidscan®(Q1000)		TAN by Titration, mgKOH/g	Optical density			Tan (delta), %
		Oxidation, Abs/mm²	ΔTAN* mgKOH/g		Dr	Dg	Db	
0	0.00	0.0	0.00	0.000	1.90	1.68	1.59	0.001
108	0.03	0.0	0.00	0.042	–	–	–	–
204	0.06	0.0	0.00	0.042	–	–	–	–
408	0.06	0.0	0.00	0.048	–	–	–	–
600	0.45	0.8	0.02	0.177	–	–	–	–
822	0.64	10.2	1.23	1.53	1.70	1.58	2.55	2.55

*ΔTAN of Fluidscan is measured in a substraction mode, i.e., that the measured value denotes the increase in the TAN of the test oil as compared to that of fresh oil

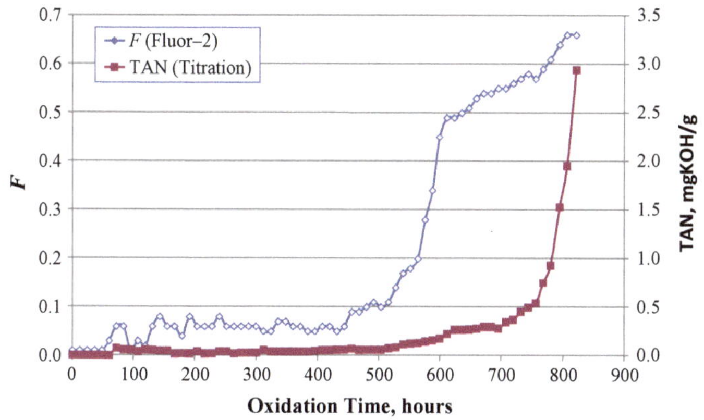

Fig. 1.19 Change of *F* parameter measured by Fluor–2 and TAN by titration method in the course of oil oxidation

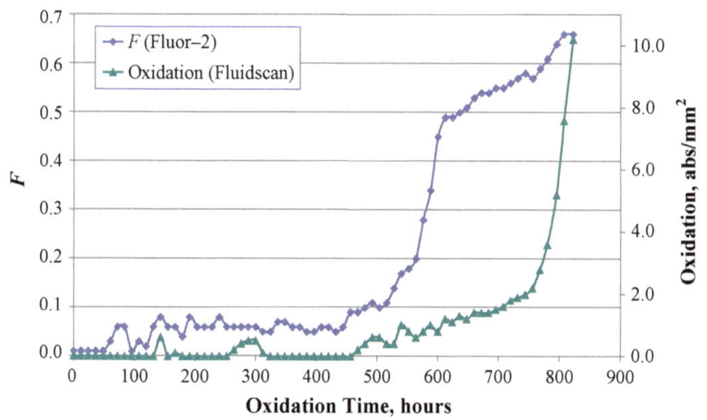

Fig. 1.20 Correlation between *F* data measured by Fluor-2 and oxidation values measured by Fluidscan

aromatic composition and polarity of the oil. It is believed that the *F* parameter shift with the oil oxidation precedes the increase in the acid number of the oil sample. This result suggests that the proposed Fluor–2 detector could be an effective indicator of the oil oxidation level.

Results showed that the newly developed apparatus used for measuring the *F* parameters had better sensitivity in quantitatively evaluating the oxidation of the

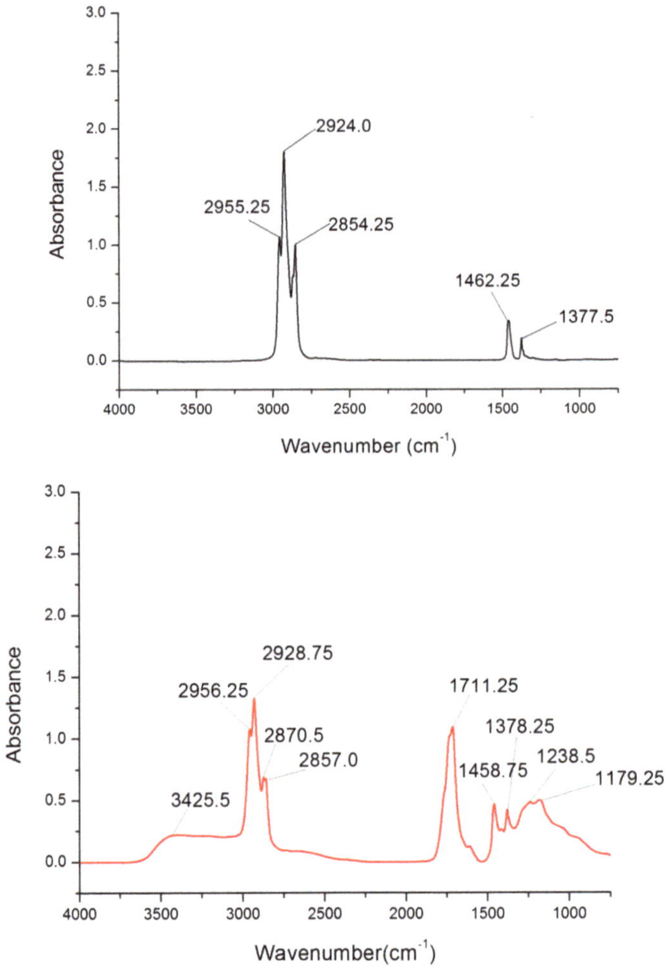

Fig. 1.21 FTIR (Perkin Elmer Co) spectra of fresh insulating oil (*top*) and FTIR spectra of oxidized insulating oil (*bottom*)

insulating transformer oil. Therefore, it is expected that the developed device could be used as a cost-effective and reliable, continuous monitoring tool for electrical insulating oils in an on-line application.

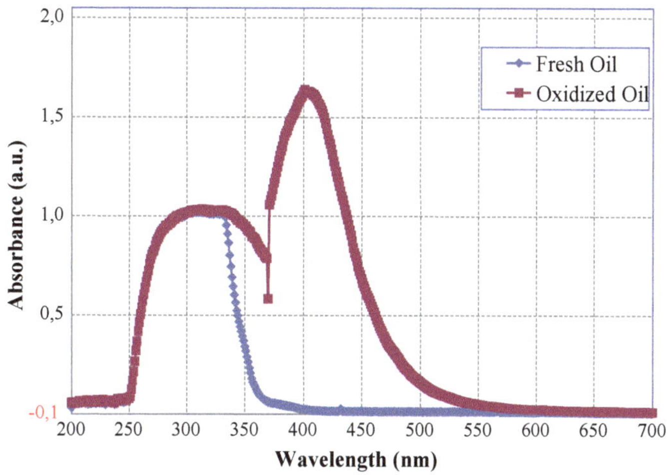

Fig. 1.22 UV–VIS spectra of fresh and oxidized insulating oil

1.6 Conclusions

Up-to-date upgraded requirements to the service characteristics of oils have initiated the development of new methods and instruments for prompt condition monitoring of lubricant. Production of new devices includes new technologies and element base, which allow improving the validity of devices being produced and reducing their cost and overall dimensions.

The choice of the diagnosis method and instrument should be governed by the type of tribosystem, its importance and cost. To assess the oil condition in automobile engines small-sized and cheap devices can be used. But in heavy machinery where more complete information on the oil condition is required, it is advisable to use integrated devices including a number of sensors. In this case we can make a valid conclusion on the oil efficiency on the basis of a complex analysis.

The fluorescent method allows for monitoring the oil oxidation in equipment operation with high reliability of the on-line diagnostics. The sensor implementing the method is distinguished by simple design, reliability, and low cost. It can be used to monitor oil condition in hydraulic systems, compressors, and turbines.

Monitoring of oil oxidation is very important for power engineering, so the techniques used in condition monitoring of tribosystems can be applied in this area. For example results obtained using the Fluor-2 detector have confirmed that this techniques could be used to analyze the transformer oil oxidation in a simple and effective manner.

References

1. L.V. Markova, N.K. Myshkin, M.S. Semenyuk, V.M. Makarenko, A.V. Kolesnikov, H. Kong, H.-G. Han, E.-S. Yoon, Methods and instruments for condition monitoring of lubricants. J. Friction Wear **24**(5), 50–59 (2003)
2. F.M. Discenzo, C.-C. Liu, D.L. Feke, L.A Dudik, US Patent 6,023,961, G01N 009/00, G01N 011/16. Micro-Viscosity sensor and lubrication analysis system employing the same—No. 09/054,117. Filed 02.04.98. Published 15.02.00
3. A.T. Pérez, M. Hadfield, Low-cost oil quality sensor based on changes in complex permittivity. Sensors **11**(11), 10675–10690 (2011). doi:10.3390/s111110675
4. S. Raadnui, S. Kleesuwan, Low-cost condition monitoring sensor for used oil analysis. Wear **259**(7), 1502–1506 (2005)
5. K.M. Park, US Patent 6297733, B60Q 001/00. Stable, Reliable Capacitive Oil Deterioration and Level Sensor—09/710,588. Filed 10.11.00. Published 02.10.01
6. K.A. Degrave, US Patent 6443006, G02F 023/00; G08B 021/00. Device Which Measures Oil Level and Dielectric Strength with a Capacitance Based Sensor Using a Ratiometric Algorithm—No. 09/567,190. Filed 09.05.00. Published 3.09.02
7. J.D. Hedges, P.J. Voelker, US Patent 7,928,741, G01R 27/26; G01N 33/26. Oil monitoring system - No. 20090201036. Filed 13.08.09. Published 19.04.11
8. S. Moon, K.K. Paek, Y.H. Lee, J.K. Kim, S.W. Kim, B.K. Ju, Multiwall carbon nanotube sensor for monitoring engine oil degradation. Electrochem. Solid-State Lett. **9**(8), H78–H80 (2006)
9. R.H. Hammerle, US Patent 5332961, G01R 027/02. Resistive Oil Quality Sensor—No. 06/927,618. Filed 06.11.86. Published 26.07.94
10. S.S. Wang, Engine oil condition sensor: method for establishing correlation with total acid number. Sens. Actuators B Chem. **86**(2–3), 122–126 (2002)
11. S.S. Wang, H.S. Lee, P.B. McGrath, D.R. Staley, US Patent 5274335, G01R 027/26; G01N 015/00. Oil Sensor Systems and Methods of Qualitatively Determining Oil Type and Condition—No. 07/863,907. Filed 06.04.92. Published 28.12.93
12. T. Takahashi, T. Kondo, US Patent 6151956, G01N 003/56; G01N 009/24; G01N 033/26; G01N 029/18. Oil Deterioration Sensor—No. 09/148,508. Filed 04.09.98. Published November 28, 2000
13. F.J. Josse, D.S. Everhart, US Patent 5852229, G01N 027/00. Piezoelectric Resonator Chemical Sensing Device—No. 08/654,993. Filed 29.05.96/ Published 22.12.98
14. F. Dickert, P. Forth, P. Lieberzeit, G. Voigt, K.D. Marquardt, US Patent 6223589, G01N 033/26. Oil Quality Sensor—No. 09/299,126; Filed 26.04.99. Published 01.05.01
15. M. Barnes, Fourier Transform Infrared Spectroscopy. Practicing Oil Analysis 3 (2002), http://www.machinerylubrication.com
16. F.R. van de Voort, J. Sedman, D. Pinchuk, An overview of progress and new developments in FTIR lubricant condition monitoring methodology. J ASTM Int. **8**(5) (2011). ID: JAI103344
17. Y. Takezawa, Y. Ito, J. Katagiri, US Patent Application 20020069021, G06F 019/00; G01N 031/00. Automobile Oil Deterioration Diagnosing Apparatus. Published 06.06.02
18. W.G. Kim, K.H. Oh, K.W. Chung, Y-W. Kim, US Patent 8752415, G01N 21/59. Method and system for measuring engine oil deterioration—No. 0130047708. Filed 28.02.13/ Published 17.06.14
19. C.M. Stellman, K.J. Ewing, F. Bucholtz, I.D. Aggarwal, Monitoring the degradation of a synthetic lubricant oil using infrared absorption, fluorescence emission and multivariate analysis: a feasibility study. Lubric. Eng. **55**(10), 42–52 (1999)
20. R.B. Thompson, Z.F. Ge, M. Patchan, C.C. Huang, C.A. Fierke, Fiber optic biosensor for Co (II) and Cu(II) based on fluorescence energy transfer with an enzyme transducer. Biosens. Bioelectron. **11**(6–7), 557–564 (1996)

21. J.B.T. Lloyd, The nature and evidential value of the luminescence of automobile engine oils and related materials. Part I. Synchronous excitation of fluorescence emission. J. Forensic Sci. Soc. 11, 83–94 (1971)
22. I.M. Warner, G.D. Christian, E.R. Davidson, J.B. Callis, Analysis of multicomponent fluorescence data. Anal. Chem. **49**(4), 564–573 (1977)
23. J.F. Fantasia, H.C. Ingrao, The development of an experimental airborne laser remote sensing system for the detection and identification of oil spills. Proceedings of the 9th international symposium On Remote Sensing of the Environment, Ann Arbor, Michegan, (1974), pp. 1711–1745
24. E.M. Hegazi, A.M. Hamdan, J.N. Mastromarino, US Patent 6,633,043, G01N 021/64. Method for characterization of petroleum oils using normalized time-resolved fluorescence spectra—No. 10/059,020. Filed 30.01.02. Published 14.10.03
25. B.W. Wilson, T.J. Peters, C.L. Shepard, J.H. Reeves, US Patent 6,810,718, G01N 011/00. Apparatus and method for fluid analysis—No. 10/339,811. Filed 10.01.03. Published 02.11.04
26. C.V. Ossia, H. Kong, L.V. Markova, N.K. Myshkin, On the use of intrinsic fluorescence emission ratio in the characterization of hydraulic oil degradation. Tribol. Int. **41**, 103–110 (2008)
27. H. Kong, E-S Yoon, H-G Han, L. Markova, M. Semenyuk, V. Makarenko, US Patent 7391035, G01N 21/64. Method and device for monitoring oil oxidation in real time by measuring fluorescence—No.11/407404. Filed 18.04.06. Published 24.06.08
28. I.A.R. Gray, A guide to transformer oil analysis, transformer chemistry services, http://www.satcs.co.za/Transformer_Oil_Analysis.pdf
29. EL-Sayed M. M. EL-Refaie, M.R. Salem, W.A. Ahmed, Prediction of the Characteristics of Transformer Oil under Different Operation Conditions. World Acad. Sci. Eng. Technol. **53**, 764–768 (2009)
30. S. Deepa, R. Sarathi, A.K. Mishra, Synchronous fluorescence and excitation emission characteristics of transformer oil ageing. Talanta **70**, 811–817 (2006)
31. B. Pradier, C. Largeau, S. Derenne et al., Chemical basis of fluorescence alteration of crude oils and kerogens: 1—microfluorimetry of an oil and its isolated fractions; relationships with chemical structure. Org. Geochem. **16**(1–3), 451–460 (1990)
32. B. Wicaksono, H. Kong, L.V. Markova, H.-G. Han, Application of fluorescence emission ratio technique for transformer oil monitoring. Measurement **46**(10), 4161–4165 (2013)

Chapter 2
Oil Viscosity Monitoring

2.1 Introduction

The safety and reliability of tribosystems depend largely on the properties of the lubricating Materials and methods of its monitoring and diagnosis [1, 2]. Viscosity is one of the major physicochemical factors of quality and efficiency of oil, its capacity to provide the effective thickness of the lubricating layer between the friction surfaces which prevents the severe wear and failure of machinery [3, 4].

Viscosity of lubricating oils depends on temperature and load-speed conditions of their operation. For example, at high speeds, low loads, and low temperatures of operation of friction units, preference is given to oils with low viscosity, while oils with high viscosity operate better at low speeds, high loads, and elevated temperatures. Thus, an important aspect of ensuring the proper operation of tribosystems is the correct selection of oil, according to viscosity and its variation with temperature. As a rule, both the selection of lubricating oils and the development of the schedule and criteria for their changing are aspects of tribosystem design.

2.2 Oil Viscosity Characterization

In accordance with the technical requirements, the viscosity-temperature properties are characterized by viscosity index. To enhance the viscosity-temperature properties, thickening additives, e.g. polymer compounds, are applied. Polymethacrylates, polyisobutenes, products of polymerization of vinilbutyl ether, and others, are used for this purpose.

It is important not only to choose the oil properly, but to maintain its viscosity during operation, which requires regular monitoring of this parameter. If a change in viscosity is detected, subsequent analysis of the oil can identify the cause of such change. Either an increase or decrease in oil viscosity can result in the disturbance

N.K. Myshkin and L.V. Markova, *On-line Condition Monitoring in Industrial Lubrication and Tribology*, Applied Condition Monitoring 8,
DOI 10.1007/978-3-319-61134-1_2

of the bearing capacity of tribosystem. Increased oil viscosity can be the evidence of its thermal destruction, oxidation, additive decomposition, or contamination with fuel, water, coolant, etc. Decreased viscosity can be caused by ingress of fuel into the oil and cracking at high temperature. At the same time, certain processes can compensate viscosity change. For example, the oil of diesel engines is contaminated both by fuel, which decreases the viscosity, and soot, which increases it [5, 6]. Anyway, variation in the oil viscosity is often the first indicator of an important problem in the tribosystem.

Oil viscosity monitoring usually involves the determination of its relative variation in the course of operation relative to the viscosity of fresh oil. It should be taken into account that the viscosity of fresh oil according to standards can differ from its nominal value up to 20%, while a change in the viscosity by 10% during operation can often be critical. Therefore, preliminary inspection of fresh oil is important for obtaining the reference value of monitoring the viscosity. Table 2.1 lists the limits of viscosity variations, which are used for the monitoring of engine and industrial oils [5].

During viscosity measurement, it is necessary to take into account that lubricating oil, depending on its composition and state, can show the properties of the Newtonian fluid in which viscosity is independent on the stress or shear rate or a non-Newtonian fluid in which the viscosity depends on these factors. Base mineral and synthetic oils show the properties of Newtonian fluid. Oxidation and contamination of oil in the course of operation result in deviation of the viscosity properties from the properties of Newtonian fluid. With the introduction of viscous additives into the base oil, which decreases the temperature dependence of the viscosity, or with the formation of a water–oil emulsion, the oil becomes non-Newtonian fluid (Table 2.2) [7].

Table 2.1 Viscosity limits [5]

Limit	Engine oil** (%)	Industrial oil** (%)	Industrial oil operating in severe loading modes** (%)
Critical (upper)	+20	+10	+7
Preventive (upper)	+10	+5	+4
Preventive (lower)	−5	−5*	−5*
Critical (lower)	−10	−10*	−10*

Notes *Value is two times higher for oils with additives improving the viscosity index
**Limits for engine oil are indicated for the viscosity at 100 °C; for industrial oil, for the viscosity at 40 °C

Table 2.2 Viscous properties of lubricating oil [7]

Oil	Viscous properties
Fresh base oil Fresh oil containing no viscous additives	Newtonian fluid
Used oil without viscous additives Average level of oil oxidation Contaminated oil	Fluid similar to Newtonian one
Highly oxidized oil Used oil with viscous additives	Fluid similar to non-Newtonian one
Fresh oil containing viscous additives Oil–water emulsion	Non-Newtonian fluid

2.3 Laboratory Measurements of Viscosity

In the case of viscosity measurement, the concepts of the dynamic and kinematic viscosity are used. The dynamic viscosity η is the ratio of shear stress between the liquid layers to the transversal gradient of velocity. The kinematic viscosity ν, i.e., the liquid resistance to flow under gravitation, is the ratio of the dynamic viscosity to liquid density ρ_l.

At present, viscosity estimation is based on the measurement of resistance to motion of a body within the medium, or by flowing the test liquid through the channel with a given geometry. Capillary, rotational, and vibration viscometry and the falling ball method are widely practiced [8].

Determination of viscosity by the capillary method is based on the Poiseuille law and consists in measurement of the time of flowing of the known amount of liquid through the capillaries with a circular section at a given pressure drop. The standard capillary method for estimation of the kinematic viscosity (ASTM D445) involves measurement of the time of outflow of the defined volume of liquid under the force of gravity through a calibrated glass capillary [9]. For all viscometers, the time of liquid outflow is proportional to its kinematic viscosity. The kinematic viscosity (ν, mm^2/s) is computed by the formula

$$\nu = Ct,$$

where C is the calibration constant for the viscometer, mm^2/s^2; t is the arithmetic mean value of the outflow time, s. The relative error for standard capillary viscometers is ±0.1–0.3%; and for operating devices, ±0.5–2.5%.

The measurement of the dynamic viscosity provides for the presence of actuated parts in the viscometer (rotational, torsion, vibration viscometry; the falling ball method, and others).

In rotational viscometers, the liquid under investigation is placed in the clearance between two coaxial bodies (cylinders, cones, and spheres) or between plane and cone. One of the bodies rotates with frequency ω_r and torque M is transmitted through the liquid to another, stationary body [10]. The dynamic viscosity η is determined by the torque at a given angular velocity or the angular velocity at a given torque by the formula

$$\eta = \frac{KM}{\omega_r},$$

where K is the constant dependent on the viscometer design.

The essence of the standard rotational method for the measurement of oil viscosity consists in recording the torque lag of an inner cylinder or cone of a gauge with the test oil at various gradients of the shear rate and in the computation of the shear stress and the dynamic viscosity:

$$\eta = \frac{\tau}{D},$$

where τ is the shear stress computed with the use of the measured relative angle of rotation of the measuring unit; D is the gradient of shear rate.

The relative error for the most widespread rotational viscometer lies within the limits of 3–5%.

In laboratory practice, the simple *falling ball method* is widespread, based on the measurement of velocity v of the steady motion of a body under the effect of gravity within the test liquid. The viscosity is computed by the Stokes formula

$$\eta = \frac{2}{9} \cdot \frac{(\rho_b - \rho_l) \cdot gR_b^2}{v_b},$$

where ρ_b and ρ_l are densities of the ball material and the test liquid, respectively; R_b is the ball radius; g is the acceleration of gravity; v_b is the ball velocity. The error of the method is ±1–3%.

Vibration viscometry is based on determination of variations in the parameters of the forced oscillations of a probe submerged into the studied medium. To measure viscosity of oil, an amplitude resonance form of the vibration method for viscometry is the most convenient. In this case, by tuning into the resonance, one can achieve maximum amplitude A of vibration; the amplitude of the vibration of the viscometer probe is the parameter used to determine the viscosity. In the general case, the dynamic viscosity is determined from the calibration relationship

$$\sqrt{\eta \rho_l} = \varphi(A),$$

where ρ_l is the oil density.

A nonstandard laboratory method of viscosity monitoring is ultrasonic viscometry based on the measurement of the dynamic viscosity by means of acoustic vibration with ultrasonic frequency. The submersion of vibrator, which performs free or forced vibration at the resonance frequency, into the liquid introduces additional losses related to the excitation of transverse acoustic waves within the liquid. This results in a decrease in the amplitude of forced vibrations or in accelerated damping of free vibrations and variation of the resonance frequency of the vibrator. The measurement of the vibration parameters is used for evaluating the liquid viscosity. Ultrasonic viscometers measure the viscosity in a range from 10^{-3} to 500 Pa s with a relative error of 5%.

To fulfill the laboratory analysis, a wide spectrum of viscometers with high accuracy has been developed; however, their great size and high cost cause difficulties for their use for on-line monitoring. Therefore, the issue of the day is the development of new effective means for viscosity monitoring in real time. If viscosity measurement devices can be mounted into the oil circulation lines, the reliability of equipment operation will be increased, and the necessity of the periodic oil sampling for laboratory analysis will be eliminated, lowering maintenance costs.

2.4 Methods of On-Line Viscosity Monitoring

Viscosity monitoring devices mounted in oil circulation lines face some particular requirements: they must continuously give out reliable information on lubricating material viscosity without the need for frequent calibration or maintenance and must have a long term of operation within the hostile environment of an operating machine at high temperatures, pressure, and vibration. At the same time, they must be compact and have low cost. In the development of on-line viscosity meters, different approaches are used: methods using the macro-displacement of a body within the liquid, and the vibration and acoustic methods (Fig. 2.1) [8].

Methods for viscosity measurement using macroscopic displacement of a body within the liquid are based on modification of the laboratory measuring procedures. The capillary method is realized through mounting of capillary with a predefined

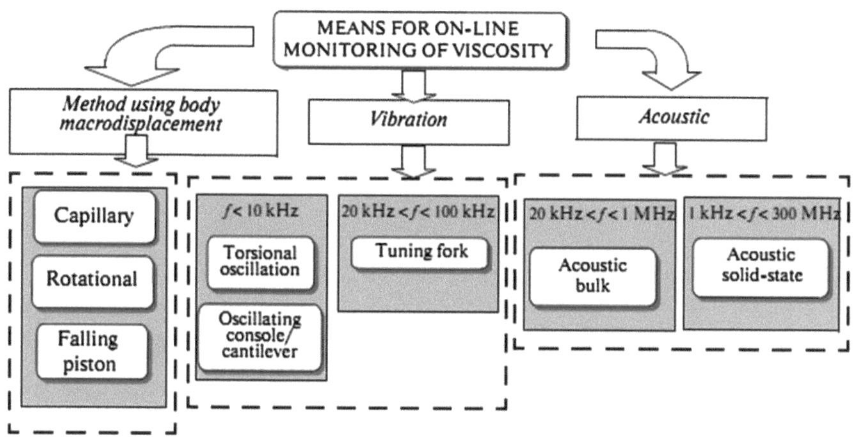

Fig. 2.1 Means for on-line monitoring of viscosity of the lubricating oils

configuration into the oil flow for measuring the pressure drop at a given flow velocity [11]. The application of this method is made difficult because the oil flow velocity is variable during machine operation and, in addition, the capillary becomes contaminated over time, decreasing measurement reliability.

Figure 2.2 presents a rotational viscometer based on Brookfield method [12], which can be mounted into the oil tank. The viscometer consists of rotating cylinder 1 fixed on the shaft 3 coaxially within motionless cylinder 2. The upper end of shaft 3 supports the measuring shaft 4 with which one end of the volute spring 5 is jointed. Its other end is fixed on the slotted disk 6 mounted on the shaft 7 of drive 9. The tested lubricating oil fills the viscometer chamber and produces viscous friction force in the oil layer within the circular clearance between the cylinders 1 and 2. The dynamic viscosity is proportional to the torque of this force, which is evaluated by the angle position of measuring shaft 4 relative to the shaft 7 of the drive. The angle position of the drive shaft is measured with slotted disk 6 and optoelectronic position sensor 8a. To determine the turning of the measuring shaft, indicator 10 and optoelectronic position sensor 8 are used.

The laboratory falling ball method was modified in the viscometer that uses the displacement of piston 1 in the liquid under the force of gravity to measure viscosity (Fig. 2.3). The piston goes up periodically by means of pneumatic mechanism 3. When the piston is in the elevated state, oil fills the measuring tube 2. Then the

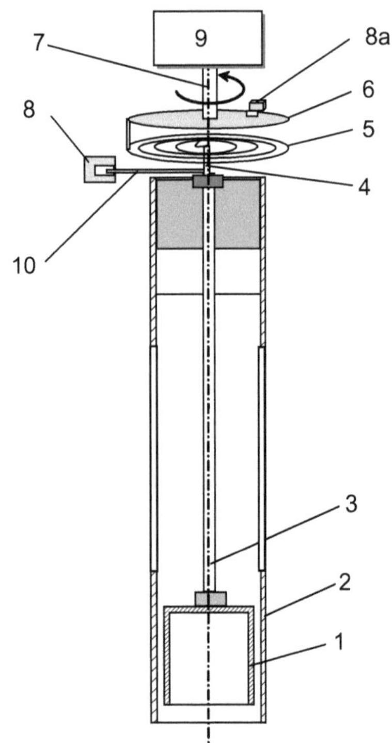

Fig. 2.2 Rotational viscometer: *1*, *2* rotational and motionless cylinders, respectively; *3* shaft; *4* measuring shaft; *5* volute spring; *6* slotted disk; *7* drive shaft; *8*, *8a* optoelectronic position sensors; *9* drive; *10* indicator [12]

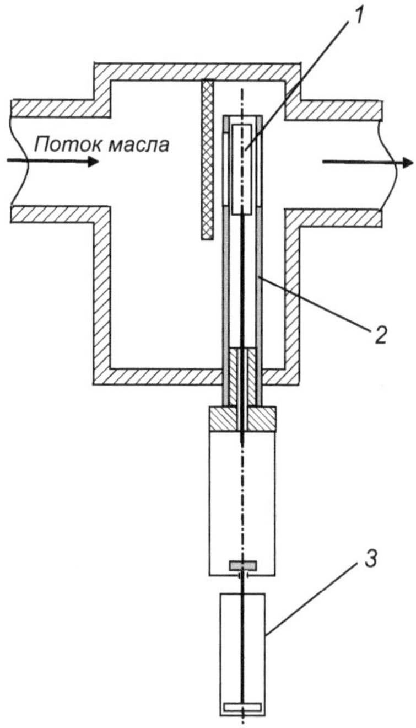

Fig. 2.3 Viscometer based on the method of the falling piston: *1* piston; *2* measuring tube; *3* pneumatic mechanism [13]

piston goes down and falls under the force of gravity, forcing out oil from the tube. The oil viscosity is evaluated by the time of the piston fall [13]. The disadvantage of the device is the necessity for mounting the piston in the strongly vertical position.

The method was perfected in a device using two electromagnetic coils 2 which enveloped the piston 1 made of ferromagnetic material (Fig. 2.4). The viscometer is mounted within the oil circulation line and oil fills the measuring chamber 4. The electromagnetic coils are switched on in turn, producing a force initiating the piston displacement to both sides along the measuring chamber. As the viscosity increases, the piston displacement becomes slower. The oil viscosity is evaluated by the time of motion of the piston between the coils. Because the piston displacement on both sides is forced, it is not subject to the effect of gravitation and oil flow [14].

The design of the viscometers based on macroscopic displacement, which are mounted into the oil circulation lines, is intricate due to their complexity, large size, and the presence of moving parts that decreases the reliability of the devices.

An alternative approach to on-line monitoring is presented by vibration and acoustic viscometers containing no moving parts subjected to wear.

Most vibration methods for on-line monitoring are based on the measurement of variation $\Delta A = A_2 - A_1$ in the amplitude of the natural oscillation of vibrator within the monitoring liquid, or of the oscillation bandwidth $\Delta f = f_1 - f_2$ [15] (Fig. 2.5). To monitor the dynamic viscosity, the phase difference for the signal exciting oscillations of vibrator and signal of its natural oscillations, as well as the rate of damping oscillations are measured.

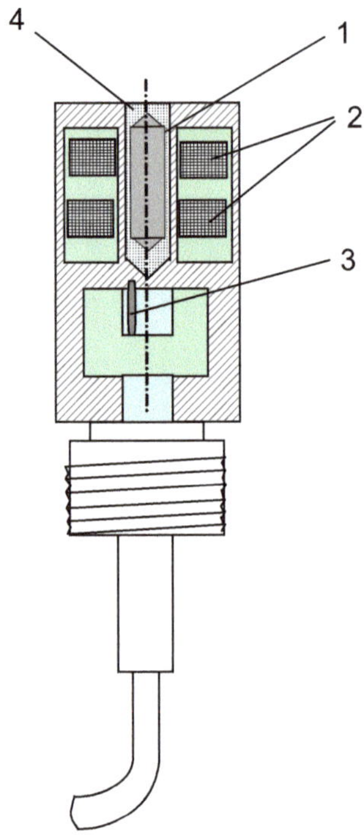

Fig. 2.4 Electromagnetic viscometer: *1* piston; *2* electromagnetic coils; *3* temperature device; *4* measuring chamber [14]

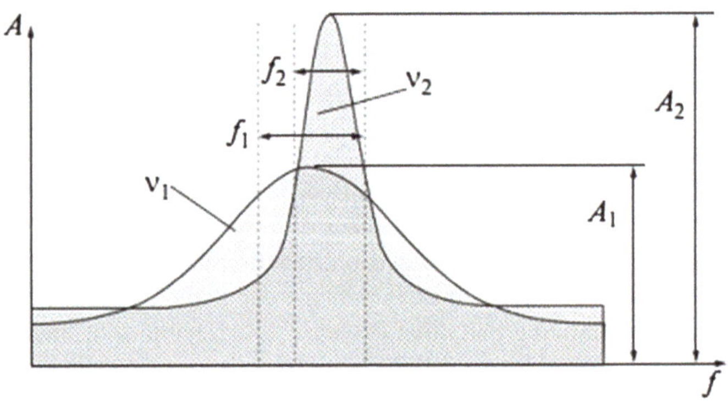

Fig. 2.5 Variation of parameters of vibrator oscillations as the liquid viscosity varies: A_1, A_2 amplitude and f_1, f_2 frequency of vibrator oscillations within the liquid with viscosity ν_1 and ν_2 ($\nu_1 > \nu_2$), respectively [15]

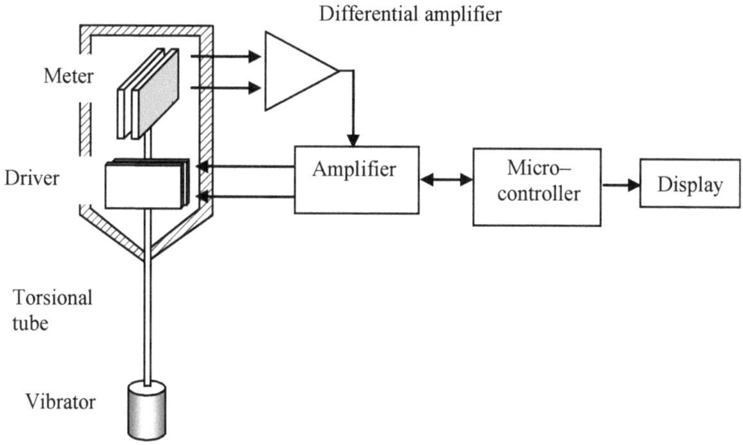

Fig. 2.6 Viscometer based on torsion oscillation [16]

Depending on the parameter used to evaluate viscosity, the device can have one of two configurations.

The viscometer based on torsion oscillations consists of vibrator (cylinder, plate, etc.) mounted at one end of the torsion tube (Fig. 2.6) [16]. At the second end of the tube, a drive and a turn meter are fixed. The drive, including two piezoelectric elements, is controlled by alternating voltage in such a way that vibrator produces torsion oscillations relative to the geometrical axis at the resonance frequency. The meter (the second pair of piezoelectric elements) records a signal characterizing the torsion oscillations. After amplification, the signal is processed with a microcontroller, whereby the phase difference between the voltage given on the piezoelectric elements of the drive and the signal from the meter, i.e. the damping factor proportional to the square root of the liquid viscosity, is determined.

The viscometer based on the use of vibrating cantilever has a beam made of a ferromagnetic material with a Teflon coating [17]. The cantilever 1 with the permanent magnet 3 is mounted over the electromagnet 2 (Fig. 2.7). The short pulse of current flowing via the electromagnet coil excites the bending oscillations of the cantilever at the frequency of its natural oscillations. Upon submergence into the liquid, oscillations are damped by the liquid resistance force, and the greater the liquid viscosity, faster they arc damping. Oscillations of vibrating cantilever at a frequency of $\sim$500 Hz are recorded by the same electromagnet coil and the dynamic viscosity of liquid is evaluated by the rate of their damping.

The disadvantage of the use of such a device is the contamination of the clearance between the cantilever and electromagnet with time and the ingress of ferromagnetic wear particles that disturb the operational efficiency of the device.

Tuning fork viscometers are based on the use of mechanical resonator usually operating at a frequency of <75 kHz [18]. The transducer includes the prongs 1 and 2 forming the tuning fork, which is submersed into the test liquid [19]. The prongs

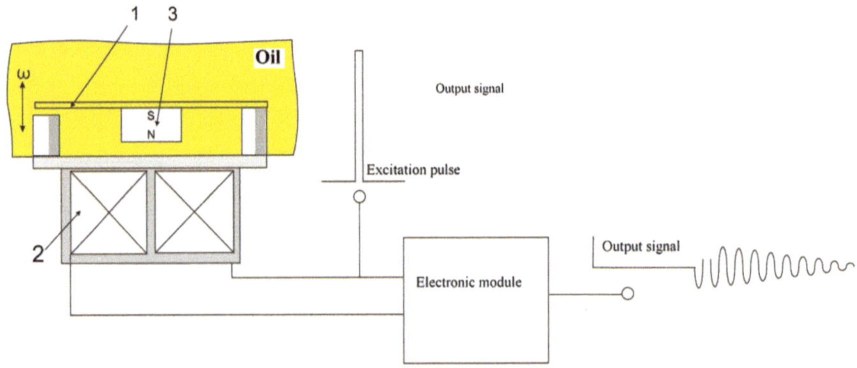

Fig. 2.7 Viscometer with bending oscillations of the cantilever: *1* cantilever; *2* electromagnet; *3* permanent magnet [17]

have channels in which piezoelectric ceramic elements 3, 4, 5, and 6 are mounted (Fig. 2.8). During the supply of voltage from the electronic module to piezoelectric elements 3 and 5, oscillations of the prongs are excited in the opposite phase. The signal describing the tuning fork oscillations is recorded by the piezoelectric elements 4 and 6. The oscillations depend on the shear resistance of the liquid and, consequently, on its viscosity. Tuning fork devices for monitoring the oil state in automotive engines are commercially produced and they can measure simultaneously the oil viscosity and degree of its contamination.

Fig. 2.8 Tuning fork viscometer: *1*, *2* prongs; *3–6* piezoelectric elements [19]

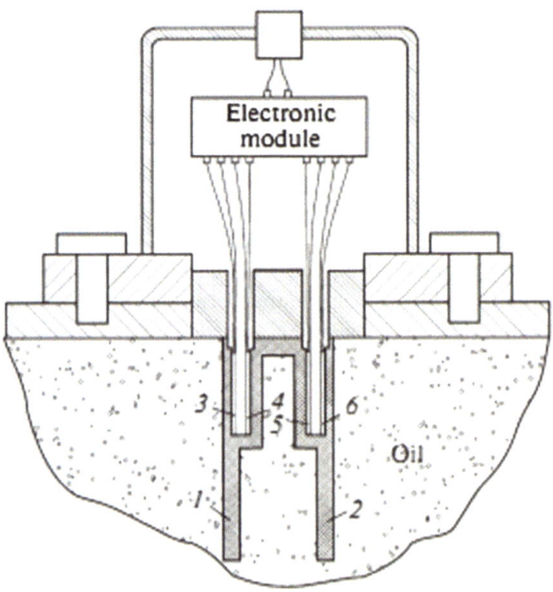

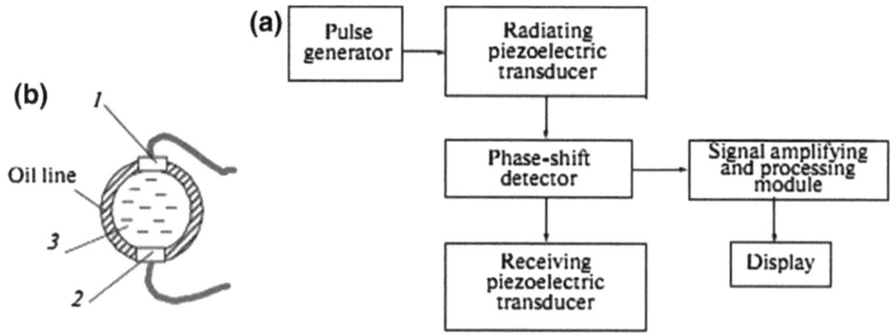

Fig. 2.9 Scheme of the acoustic viscometer **a** and transducers in the oil line **b**: *1*, *2* radiating and receiving piezoelectric transducers, respectively; *3* test oil [20]

The principle of operation of ***acoustic viscometers*** is based on the excitation of elastic compression waves within the liquid.

Figure 2.9a shows the schematic of bulk acoustic viscometers including pulse generator, radiating and receiving piezoelectric transducers, and detector of the phase shift [20]. To measure viscosity the radiating 1 and receiving 2 piezoelectric transducers are mounted in the walls of the oil line (Fig. 2.9b). One of the sides of the radiating piezoelectric transducer is in contact with the test liquid. Pulses of the generator with a given frequency induce oscillations of the radiating piezoelectric transducer and excite the longitudinal acoustic wave within the test oil. The receiving piezoelectric transducer is opposite to the oscillator at a given distance and records the acoustic wave propagating through the liquid. The acoustic wave carries the information on the time of relaxation of liquid molecules, which depends on the molecular structure of the liquid and is expressed in the phase difference of the wave excited by the oscillator and the wave passed through the liquid and recorded by the receiver. The phase difference measured by the detector is used to determine the wave velocity, and the kinematic viscosity is computed according to the formula

$$\nu = \frac{V^2}{2\omega},$$

where V and ω are velocity and frequency of the acoustic wave, respectively.

The optimum operating frequency for acoustic viscometers lies in the range of 35–45 kHz. The main disadvantage of such devices is that the acoustic wave velocity depends not only on the viscosity of oil, but also on the level of its contamination by wear particles, air bubbles, etc.

Solid-state viscometers based on the use of acoustic waves present a promising class of devices [21, 22]. The physical principle of their operation is energy transfer by the acoustic shear wave from a waveguide having characteristic impedance of the material $Z_w = (\rho_w \mu)^{1/2}$ into the adjacent layer of the test liquid with wave resistance $Z_l = (\omega \rho_l \eta)^{1/2}$, where ρ_w and ρ_l are the densities of the waveguide and the liquid, respectively and μ is the elastic modulus of the waveguide at shear. The energy transfer is proportional to the ratio Z_l/Z_w under condition $Z_l \ll Z_w$. In the application of solid-state viscometers, the notion of "acoustic viscosity" (a.v.) is used, that is the product of dynamic viscosity and density of liquid: a.v. = $\eta\rho_l$.

In viscometers with acoustic waves, piezoelectric materials are used to generate acoustic waves by an electric field. The waves are propagated through or over the surface of the substrate and then transformed back into the electric field for measuring. All variations in the characteristics of the propagation path affect velocity and/or amplitude of the wave which relate to the properties of the controlled medium.

Solid-state viscometers are classified by the mode of the wave propagating through the piezoelectric substrate. The fundamental modes of oscillation applied in acoustic wave devices for measuring liquid viscosity are thickness shear mode (TSM), shear-horizontal acoustic plate waves (SH–APW), and Love waves. These waves radiate an insignificant part of their energy into the liquid, which allows for devices to operate without substantial absorption in the liquid. The TSM and SH–APW waves are bulk, i.e., waves propagating within the entire volume of the substrate [21].

Devices based on TSM waves are the most simple and widespread. As is shown in Fig. 2.10a, such a device consists of the thin quartz disk with two circular electrodes deposited on the face surfaces. The application of voltage between the electrodes produces shear deformation of the crystal. The oscillating surface generates a laminar flow within the contacting layer of the liquid. The liquid layer involved in motion experiences delay by the phase which increases as the distance from the crystal surface rises. If the frequency of the applied voltage is varied, one can obtain the mechanical resonance and then half of the length of the acoustic wave is equal to the crystal thickness. In the case of the resonance, the maximum shear within the crystal is at its surfaces, and, consequently, the liquid viscosity exerts the greatest influence on the parameters of the crystal resonance. The shift of the resonance frequency and the change of the oscillation amplitude give information on the oil properties [22, 23]. Devices for the monitoring of engine oil viscosity based on TSM waves with resonance frequencies of 5–6 MHz are described elsewhere [24, 25]. The TSM viscometer is remarkable for its simplicity of production, resistance to harsh ambient conditions, and thermal stability. However, it was found that the output signal depends not only on viscosity, but also on the oil oxidation rate [25].

In the viscosity devices based on SH–APW waves, they use a thin piezoelectric plate with two interdigital transducers (IDTs) deposited at the opposite ends of one of the plate surfaces (Fig. 2.10b) [21]. The application of alternating voltage to IDTs causes the displacement of particles of the plate material exciting the acoustic wave. The crystal cut was selected in such a way that the application of electric field to IDTs generates shear waves with shear-horizontal polarization, so the direction of particles shift is normal to the direction of wave propagation and parallel to the sensitive surface. The wave is propagated via the plate, and on the IDTs of the other

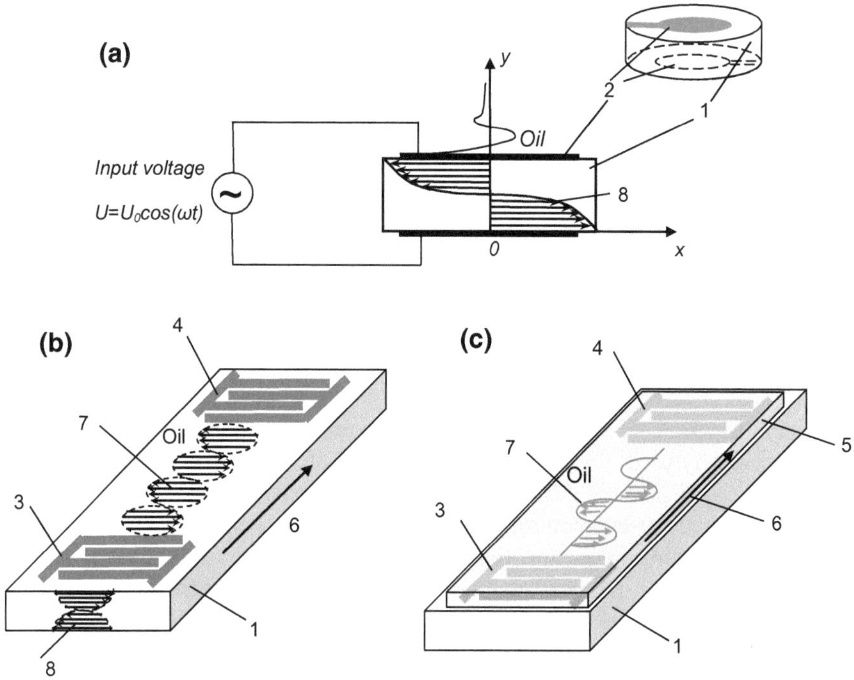

Fig. 2.10 Solid-state acoustic viscometers based on volume TSM [21] **a** and SH–APW waves [21] **b** and surface Love waves [27] **c**: *1* piezoelectric plate; *2* electrodes; *3*, *4* input and output IDTs; *5* waveguide layer; *6* direction of wave propagation; *7* surface displacement; *8* volume displacement

end of the plate the mechanical energy of the acoustic wave is transformed into the electrical one. The operating frequency of the viscosity device is 25–200 MHz. The use of such a viscometer to monitoring the lubricating oil viscosity is described elsewhere [22, 26].

Devices based on the use of surface acoustic waves with horizontal shear polarization (SH–SAW), in which the displacement of particles of the plate material occurs within the plane of the crystal surface, are also known (Fig. 2.10c) [27]. In the common case, it is impossible to obtain purely horizontal shear waves; a part of the energy is lost with the volume acoustic wave which is propagated perpendicular to the surface. To decrease the energy loss, a special layer is deposited at the surface of the piezoelectric plate with IDTs. The speed of the shear wave propagation within this waveguide layer is lower than within the plate; as a result, the acoustic energy is concentrated within this layer. Thus, conditions arise for the propagation of so-called Love waves via the layer. The sensitivity of such acoustic waves to processes occurring at the surface is higher than in the case of common SH–SAW oscillations. In spite of the small sizes and the implementation in the form of chips, viscometers based on acoustic waves have a high cost because of the expensive materials used for production of the acoustic waveguides. In addition, the high operating frequencies (1–200 MHz) require the use of complicated electronics. The devices are sensitive only to properties of thin layer of oil at the interface area because the depth of penetration of the acoustic wave into the liquid is inversely proportional to the square of frequency. The use of acoustic devices is limited to high viscosities because at oscillation with high frequencies, high molecular liquids have the tendency to behave as gels since the oscillation frequencies of molecules of such liquids are lower than the oscillation frequencies of the device.

2.5 Viscosity Sensor Based on Magneto-Elasticity

The main drawback of solid-state acoustic viscometers is that they are too expensive to be widely used; this is due to the necessity of using the expensive materials to fabricate acoustic waveguides and relatively expensive electronic components necessary to generate and control the high operating frequency. In addition, owing to the high operating frequency, these devices are sensitive to the fluid properties only in the thin layer that contacts the surface of the sensor. When high-frequency elastic waves propagate in high-molecular fluids, the latter start to show the behavior of gels and the gage readings may reflect inadequately the viscosity properties of the oil.

In parallel with solid-state sensors based on acoustic waves, magnetoelastic viscosity sensors are currently under development; they operate at lower frequencies.

2.5.1 Magnetoelastic Viscometry Technique

The magnetoelastic viscometry method [28] is based on the mutual effect of the magnetization and elastic deformations of the medium, which appears as variations in the dimensions and shape of the specimen when it is magnetized (magnetostriction) and as changes in the magnetization of the specimen when it is deformed (the magnetoelastic effect or Villary effect). The method involves the excitation of a longitudinal standing elastic wave in a sensitive element by generating an alternating magnetic field in the zone of its location with a frequency equal to the frequency of the natural oscillations of the element.

The theoretical model of the magnetoelastic viscosity sensor is based on the equation of longitudinal oscillations of thin elastic plate affected by the alternating magnetic field and the friction force that is applied to the plate immersed in the viscous liquid. For a plane wave, which propagates along the Y axis, the equation that describes the longitudinal oscillations of the plate lying in the YZ plane and oscillating in the direction of the Y axis is

$$\rho_{pl}\frac{\partial^2 u_y}{\partial t^2} = \frac{E_{pl}}{1-\sigma_{pl}^2}\frac{\partial^2 u_y}{\partial y^2} \tag{2.1}$$

where ρ_{pl} is the plate material density; E_{pl} and σ_{pl} are the Young's modulus and Poisson's ratio of the plate material, respectively; and u_y is the component of displacement vector of the plate particles along the Y axis under the effect of the external force that is applied per unit of the plate surface area. In addition to longitudinal waves that are excited by oscillations of the plate in the direction of the Y axis, rapidly damping transversal or shear waves exist in the viscous liquid, propagating along the X axis; the particles in the waves move along the Y axis (Fig. 2.11).

The equation of the motion of a viscous liquid in the vicinity of the plate surface, or the Navier-Stokes equation, can be presented in the form of the diffusion equation:

$$\frac{\partial \upsilon}{\partial t} = \frac{\eta}{\rho_l}\frac{\partial^2 \upsilon}{\partial x^2} \tag{2.2}$$

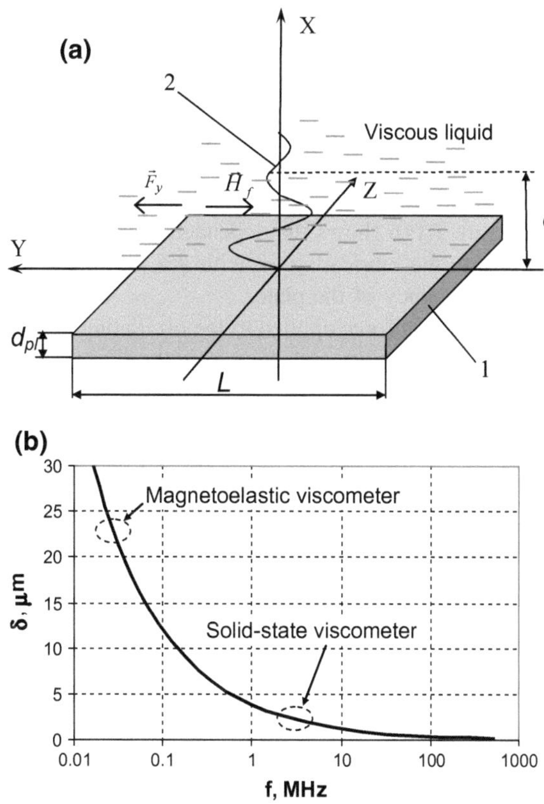

Fig. 2.11 Magnetoelastic plate under the effect of AC magnetic field $\vec{H}_f$ and friction viscous force $\vec{F}_y$ **a**: *1* plate, *2* shear wave damped by viscous liquid; dependence of shear wave penetration depth δ in a liquid on operation frequency **b**

where v is the liquid velocity; η is the liquid dynamic viscosity; and ρ_l is the liquid density.

The liquid velocity in the vicinity of the oscillating plate can be written as the following periodic function:

$$\upsilon = s_0 e^{i(kx-\omega t)} = se^{ikx} \tag{2.3}$$

Based on Eqs. (2.2) and (2.3), we derived the following dependence of the liquid velocity on the distance to the surface of the oscillating plate (Fig. 2.11):

$$\upsilon = s\frac{\sin\left[k\left(d_{pl}-x\right)\right]}{\sin(kx)},$$

where $k = \frac{1+i}{\delta}$, d_{pl} is the plate thickness; and δ is the depth of the penetration of excitation induced by plate oscillations into the liquid, which can be estimated from the following equation:

$$\delta = \left(\frac{\eta}{\pi\rho_l f}\right)^{1/2}, \tag{2.4}$$

where η is the liquid dynamic viscosity, ρ_l is the liquid density, and f is the oscillation frequency of the plate.

When moving away from the oscillating plate, the amplitude of transversal waves damps according to the exponential law; the depth of the penetration of oscillations into the liquid diminishes with increasing frequency and rises with increasing liquid viscosity.

Figure 2.11b shows the dependence of the depth of the penetration of the shear wave into the lubricating oil with η = 40 mPa s and ρ_l = 870 kg/m^3 on the oscillation frequency of the plate.

The friction force applied to the oscillating plate is directed along the Y axis and depends on viscosity and changes in the shear rate on the plate surface according to the Newton law:

$$F_y = \eta\left(\frac{\partial \upsilon}{\partial x}\right)_{x=0} \tag{2.5}$$

Equation (2.1), describing the longitudinal oscillations of the plate with account for the friction force becomes

$$\rho_l\frac{\partial^2 u_y}{\partial t^2} = \frac{E_{pl}}{1-\sigma_{pl}^2}\frac{\partial^2 u_y}{\partial y^2} - F_y \tag{2.6}$$

A standing wave appears in the oscillating plate; when one edge of the plate is fastened, the equation of the wave is

$$u_y = 2Pe^{-i\omega_n t} \cos \frac{n\pi y}{2L} \tag{2.7}$$

where P is the amplitude of the standing wave and L is the plate length.

Using the Eqs. (2.1)–(2.7), we derive the following relation for estimating the natural frequency f_a of oscillations of the plate in nonviscous medium or air:

$$f_a = \frac{n}{4L}\sqrt{\frac{E_{pl}}{\rho_{pl}(1-\sigma_{pl}^2)}}, \quad n = 1, 2, 3, \ldots \tag{2.8}$$

Thus, the natural frequency of the magnetoelastic plate depends on its dimensions and the material properties.

Similarly, using the Eqs. (2.6) and (2.7), one may estimate the natural frequency of oscillations of the plate in a viscous liquid with account for the rheological properties of the latter.

The natural oscillations of the plate immersed in a viscous liquid located in the YZ plane and oscillating in the direction of the Y axis are harmonic oscillations with exponentially damping amplitude (Fig. 2.12); they are described by the following equation:

$$y(t) = Ae^{-\xi\omega_a t}\cos(\omega_l t + \varphi), \tag{2.9}$$

where A is the oscillation amplitude; ξ is the damping viscosity factor; ω_a is the circular frequency of natural oscillations of the plate in air (without friction); ω_l is

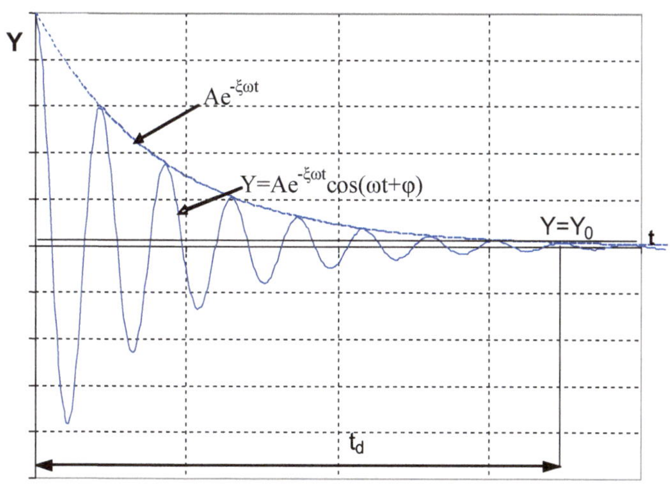

Fig. 2.12 Longitudinal oscillations of plate immersed in liquid

the frequency of natural oscillations of the plate in the liquid, $\omega_l = \omega_a\sqrt{1-\xi^2}$ and φ is the phase shift.

The rate of decay of the longitudinal oscillations of the plate immersed in the viscous liquid is described by the damping factor ξ, which depends on the liquid viscosity. When the plate is immersed in the liquid, the frequency f_l of its natural oscillations or the resonance frequency diminishes by Δf:

$$\Delta f = f_a - f_l = \frac{\sqrt{\pi f_a}}{2\sqrt{2}\pi\rho_{pl}d_{pl}}(\eta\rho_l)^{1/2} \tag{2.10}$$

Thus, when the frequency shift Δf is measured, the "acoustic viscosity," *a.v.*, or the product of the oil viscosity and oil density can be found as follows:

$$a.v. = \eta\rho_l = \Delta f^2\frac{8\pi\rho_{pl}^2d^2}{f_a} = B\frac{\Delta f^2}{f_a} \tag{2.11}$$

where B is a constant that depends on the density and thickness of the magnetoelastic plate.

Compared to viscosity, the oil density varies only slightly during operation. In this case, the following relation can be used to estimate the dynamic viscosity:

$$\eta = \frac{a.v.}{\rho_l} = B\frac{\Delta f^2}{f_a\rho_l} \tag{2.12}$$

Among the important problems that arise when developing magnetoelastic instrumentation for viscosity monitoring is the temperature dependence of the frequency of natural oscillations of the plate. It results from the temperature dependences ρ_{pl} (T) of the plate material density and $L(T)$, $d(T)$ of the plate dimensions. Thus, temperature compensation has to be provided when developing methods for measuring the oil viscosity.

In addition to estimation of the rheological properties of liquids by the difference of the frequencies of natural oscillations of the plate in air and in the liquid being monitored, acoustic solid-state sensors involve a method based on analysis of the damping rate of the oscillation amplitude, i.e., measurement of the damping factor [29, 30]. Apparently, this method can also be applied in magnetoelastic sensors for on-line diagnostics.

2.5.2 *Magnetoelastic Viscometer*

Sensitive elements made of amorphous metals, so-called metallic glasses, are promising for application in magnetoelastic sensors for on-line monitoring. These materials are alloys with an amorphous structure that results from very rapid cooling of a melt [31].

The relatively cheap amorphous ferromagnetic alloys $Fe_{40}Ni_{38}Mo_4B_{18}$ (Metglas 2826 MB) and $Fe_{81}B_{13.5}Si_{3.5}C_2$ (Metglas 2605SC) with the high ultimate strength (1000–1700 MPa) are the most widely used in miniature sensors. In addition, they possess magnetostriction (relative elongation $\lambda = \Delta L/L$ under the effect of magnetic field) of the order of 10^{-5} and high coefficient of transformation of magnetic energy into elastic one of 0.98, whereas such coefficient of most widely applied nickel alloys is $\approx$0.4. The use of this material allows designers to make sensors more portable.

It should be noted that the elasticity of these materials depends on the strength of the applied magnetic field. In order to optimize the selection of the operating point of the sensitive element, i.e., to obtain the maximal change in the relative elongation under the effect of alternating field, a stationary bias magnetic field is additionally applied. Figure 2.13 shows the dependence of the magnetostriction λ on the strength H of the stationary bias magnetic field [32]. It is seen that the use of a stationary bias field with the strength H_0 and the excitation of oscillations by an alternating magnetic field with the amplitude H_f yield variations in the relative elongation of the sensitive element equal to $\Delta\lambda$. The strength of the applied magnetic field apparently governs the frequency and amplitude of the natural oscillations of the sensitive element (Fig. 2.14) [33].

Researchers from the V.A. Belyi Metal–Polymer Research Institute of NAS of Belarus and the Korea Institute of Science and Technology have developed a magnetoelastic viscometer (Fig. 2.15) that comprises a probe and electronic unit.

The viscometer probe consists of a viscosity transmitter–sensitive magnetoelastic plate (1) 37 × 6 × 0.03 mm in size made of amorphous metallic glass Metglas 2826 MB—and an electromagnetic coil (2) (Fig. 2.15a). The latter serves to generate signal that excites mechanical oscillations of the plate and to measure the signal induced by the plate oscillations, which attenuate as they are damped by viscous oil (4). A platinum thermometer (3) is used to measure the oil temperature,

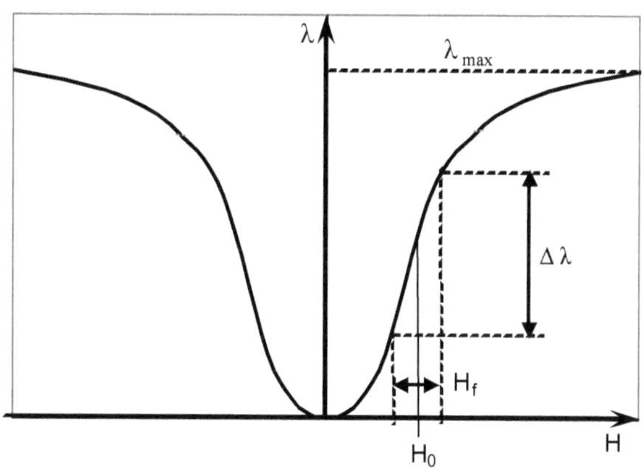

Fig. 2.13 Dependence of magnetostriction on strength of bias magnetic field

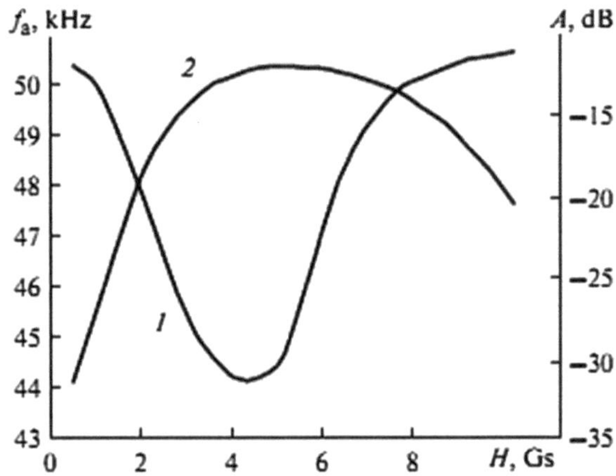

Fig. 2.14 Dependence of the resonance frequency *1* and amplitude of natural oscillations *2* of the sensitive magnetoelastic element on the strength of the stationary bias field [33]

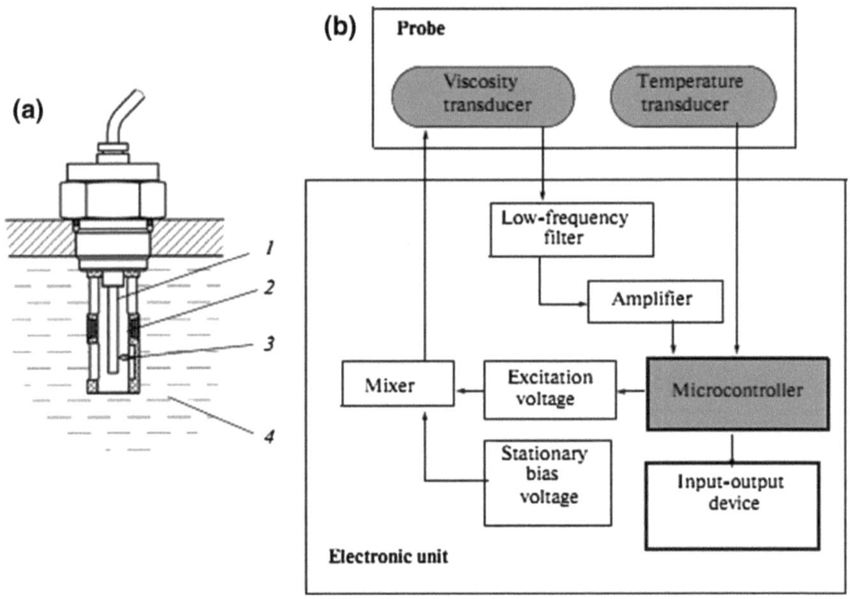

Fig. 2.15 Design of the probe **a** and block diagram of the viscometer **b** of magnetoelastic viscometer: *1* magnetoelastic plate, *2* electromagnetic coil, *3* thermometer, *4* oil

which is important when interpreting viscosity data, and to perform temperature compensation. The probe case has holes through which oil flows to the thermometer (3) and magnetoelastic plate (1). To protect the magnetoelastic element against mechanical damage end of the case that is immersed in the oil is equipped with protective mesh. Using the threaded joint and gasket ring, it is possible to mount the probe into a tank containing the oil being monitored or into a pipe through which the oil is pumped.

The determination of viscosity is based on finding the resonance of the oscillations of the plate immersed into the monitored oil and analysis of the characteristics of the signal. The viscometer implements both the method for determining the viscosity by variations in the resonance frequency and the method based on the analysis of the decay curve of the oscillation amplitude.

The program implements the following measurement algorithm. A direct current passes the turns of an electromagnetic coil (2) and induces a stationary magnetic field that magnetizes the elastic element (1), thus providing conditions for its effective operation. Simultaneously, alternating current passes the coil turns; it induces an alternating magnetic field that excites longitudinal elastic oscillations of the magnetoelastic element. Then the alternating current is switched off and the element continues oscillating at its natural frequency, which depends on the oil viscosity. The elastic oscillations of the element (1) generate alternating current in the coil (2); its frequency f_l corresponds to the frequency of oscillations of the element.

To determine the frequency f_l the frequency of the inducing alternating current has to be varied or scanned within the $f_{l,min}$–$f_{l,max}$ range. The edge frequencies of this range are found from the limiting values η_{min} and η_{max} of the monitored viscosity of lubricating oils using the formula (2.10):

$$f_{l,\min}(\eta_{\max}) = f_a - \Delta f(\eta_{\max}) = f_a - \frac{\sqrt{\pi f_a}}{2\sqrt{2}\pi\rho_{pl}d_{pl}}(\eta_{\max}\rho_l)^{1/2},$$

$$f_{l,\max}(\eta_{\min}) = f_a - \Delta f(\eta_{\min}) = f_a - \frac{\sqrt{\pi f_a}}{2\sqrt{2}\pi\rho_{pl}d_{pl}}(\eta_{\min}\rho_l)^{1/2}.$$

When using the method of determining viscosity by variations in the resonance frequency, the frequency f_{ex} is recorded, at which the resonance occurs, i.e., $f_l = f_{ex}$. The compensation of the temperature dependence of the frequency of the natural oscillations of the magnetoelastic element involves the determination of the frequency $f_a(T)$ of its natural oscillations in air from the analytical dependence stored in a microcontroller. This value of the frequency corresponds to the measured temperature T and is used to correct the shift of the frequency $f_l(T)$ of the damped oscillations in the oil:

$$\Delta f_l(T) = f_a(T) - f_l(T) \qquad (2.13)$$

The acoustic viscosity *a.v.* is calculated using the formula (2.11).

The method based on the analysis of the decay curve of the oscillation amplitude involves finding the time t_d for damping of oscillations at the resonance frequency to stop (Fig. 2.12). In this case the resonance frequency depends on the frequency at which the greatest number of pulses occurs. The damping time is governed by the number of oscillations of the plate or the number of pulses recorded with amplitude exceeding the preset limit Y_0. The viscosity is estimated from the preregistered calibration dependence $a.v.(N)$. Data are measured and processed by a microcontroller C8051F021. The display of the electronic unit shows the oil viscosity and the temperature at which the viscosity has been measured.

2.5.3 Experimental Results

Verification of data measured by two developed techniques

We have analyzed the theoretical and rated temperature dependencies of kinematic viscosity of synthetic oils. The theoretical dependencies were found in accordance with ASTM D341 for the liquids with above 2 cSt viscosity using the formula [34]:

$$\lg(\lg(v + 0.7)) = a - b \cdot \lg(T + 273), \quad (2.14)$$

where T—oil temperature, °C; ν—kinematic viscosity, cSt; *a*, *b*—coefficients found from kinematic viscosity values at two specified temperatures (40 and 100 °C).

The experimental data on kinematic viscosity were obtained by recalculation of the measured acoustic viscosity as follows:

$$v = \frac{a.v.}{\rho_l^2}. \quad (2.15)$$

Figure 2.16 illustrates the theoretical and experimental temperature dependencies of the kinematic viscosity obtained by the technique based on the resonant frequency shift (data for synthetic oil PAO4 are presented in Fig. 2.16a) and by the analysis of the amplitude decay curve (for synthetic oil PAO8, Fig. 2.16b).

The results have proved that the technique of viscosity measurement based on the oscillations decay rate yields more accurate data (±5% error) than that based on the resonant frequency shift (±15% error).

Estimation of correlation between the readings of magnetoelastic, capillary and solid-state acoustic viscometers in testing base synthetic and mineral oils

Viscosities of four synthetic (PAO4, PAO6, PAO9 and PAO40) and two mineral base oils (P-96 and P-480) were evaluated. Table 2.3 presents the kinematic viscosity values measured at 40 °C by capillary viscometer Cannon-Fenske Viscometer (Glass Capillary Viscometer, ASTM D446) following the standard method (ASTM D445), acoustic viscosity values measured by magnetoelastic

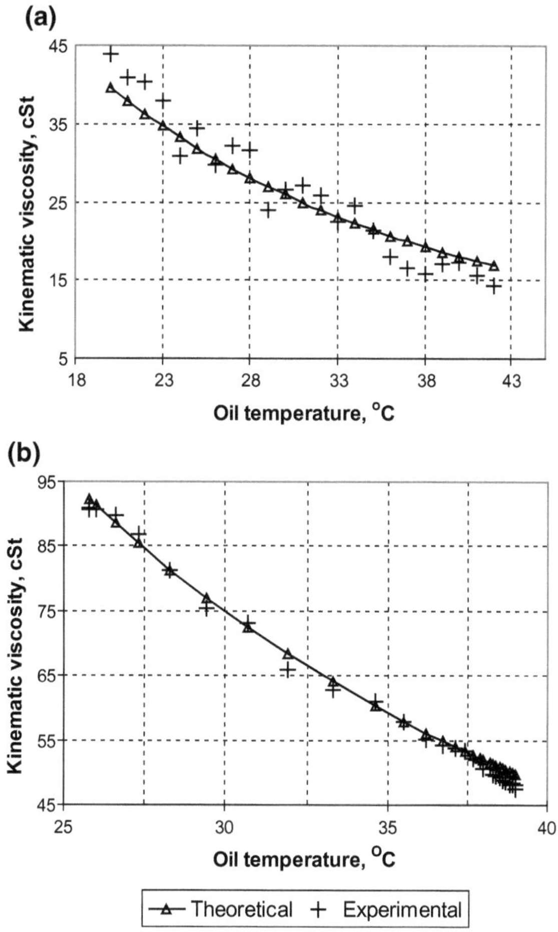

Fig. 2.16 Theoretical and experimental temperature dependencies of kinematic viscosity using the technique of viscosity estimation from the resonant frequency variation **a** and the analysis of the amplitude decay curve **b**

Table 2.3 Measurement results of oil viscosity values by capillary, magnetoelastic and solid-state viscometers at 40 °C

Viscometer	Measured parameter	Oils under test					
		PAO 4	PAO 6	PAO 8	P-96	PAO 40	P-480
Capillary viscometer	ν, cSt	17.3	30.3	47.7	96.2	392.6	496.6
Magnetoelastic viscometer	a.v., cP kg/cm^3	12.7	21.8	30.0	70.0	299.6	417.6
	ν, cSt	19.3	32.6	44.2	92.1	423.6	533.2
Solid-state viscometer	a.v, cP kg/cm^3	12.8	20.2	31.2	79.2	310.0	468.2
	ν, cSt t	19.5	30.2	45.9	104.2	438.3	597.8
ASTM D 891-09	ρ_l, g/cm^3	0.811	0.818	0.824	0.872	0.841	0.885

viscometer using the procedure of analyzing the decay rate of oscillations, and by solid-state acoustic viscometer (ViSmart, Vectron Int.). The table also lists the corresponding values of kinematic viscosity calculated by formula (2.15) with account of oil density measured under ASTM D891-09.

The comparison of measurement results shows that the data obtained by the developed viscometer are the closest to viscosity measured by the capillary technique within the whole measurement range 17–500 cSt. In contrast, the solid-state acoustic viscometer has shown less validity in measuring viscosity above 300 cSt. Its readings are much higher the expected ones, evidently because the layer of viscous oil on the sensing surface of the viscometer behaves like a gel under high operating frequencies (5.3 MHz).

Changes in oil viscosity during artificial aging

Oil samples were prepared by artificial aging. The gear oil GS Caltex "Meropa 220" was oxidized at temperature 148 °C. The oxidation process was stopped for the nighttime. In the course of oil oxidation, the samples were taken after 0, 8, 26 and 35 h. Viscosity was measured using the magnetoelastic, solid-state, and glass capillary viscometers at 40 °C (Table 2.4).

The comparison of the test results proves that the solid-state viscometer shows higher viscosity values than those measured by the magnetoelastic and capillary viscometers. This can be attributed to variations in chemical structure during artificial aging and increased polarity of molecules, which changes interfacial interactions between the oil and sensitive surface of the solid-state viscometer. Since the viscometer is sensitive to variations in properties within a very thin oil layer, these interactions at the interface can affect the readings.

Effect of PMMA viscosity index improver on viscometer readings

To study the effect of PMMA VI improver on viscometer readings we have prepared the samples of synthetic base oil PAO6 with improver in concentrations 3, 6

Table 2.4 Viscosity monitoring in the course of oil aging

Oil sample code	Sample#1	Sample#2	Sample#3	Sample#4
Duration of oxidation, Hours	0	8	26	35
Days	0	1	3	4
Capillary viscometer ASTM D 446, ν, cSt	236	239	245	247
Solid-State viscometer a.ν, cP g/cm^3	201	205	210	212
ν, cSt	248	253	259	262
Magnetoelastic viscometer a.ν, cP g/cm^3	197	198	199	204
ν, cSt	243	244	246	251

Table 2.5 Kinematic viscosity of oil samples with VI improvers

VI improver	M_W[a], Daltons	Kinematic viscosity, cSt at 40 °C			
		C_m[b] = 0%	C_m = 3%	C_m = 6%	C_m = 9%
PMMA-1	40,311	30.40	34.18	38.16	42.62
PMMA-2	60,795	30.40	33.40	37.14	41.78
PMMA-3	155,259	30.40	34.40	39.04	44.55
PMMA-4	530,537	30.40	33.31	37.16	42.42
kinematic viscosity, cSt at 100 °C					
PMMA-1	40,311	5.98	6.68	7.54	8.47
PMMA-2	60,795	5.98	6.68	7.54	8.60
PMMA-3	155,259	5.98	7.04	8.34	9.94
PMMA-4	530,537	5.98	7.15	8.67	10.50

[a]Molecular weight
[b]Concentration of VI improver

and 9 wt%. Four different modifications of PMMA (PMMA-1, PMMA-2, PMMA-3 and PMMA-4) with different molecular weights were under testing to estimate the influence of molecular weights of the VI improver on viscometer data (Table 2.5). The molecular weights of the VI improvers were measured using GPC (Gel Permeation Chromatography, HLC-8320GPC) at 40 °C temperature.

The improver molecular weight effect on kinematic viscosity at 40 and 100 °C was estimated by Capillary Viscometer Cannon Fenske (ASTM D445) (Table 2.5). As it was expected, kinematic viscosity increased at 100 °C with increasing molecular weight and content of the VI improver. Meanwhile, at 40 °C viscosity increased with growing concentration of the VI additive but without any correlation between the increment in molecular weight and viscosity. Obviously, the VI improvers have different temperature behaviors, which is evident at lower temperature (40 °C), while their application is intended for higher temperature (100 °C).

Viscosity of the test samples was also measured by the magnetoelastic and solid-state viscometers at 40 °C. The absolute viscosity was determined from the measured acoustic viscosity values using formula

$$\eta = \frac{a.v.}{\rho_l}.$$

The analysis of measurement results of the absolute viscosity of the test oils with addition of PMMA-1 and PMMA-2 in concentrations 3, 6 and 9 wt% shows that the data determined using the capillary viscometer coincide with the readings measured by the magnetoelastic viscometer (Fig. 2.17a, b). This fact proves that the test oil samples preserve the properties of the Newtonian fluid at the operating frequency < 30 kHz (at shear rate $<10^3$ s^{-1}). Viscosity values of these oils measured by the solid-state viscometer are much lower than that obtained by the capillary and magnetoelastic viscometers. This confirms that the oils behave as

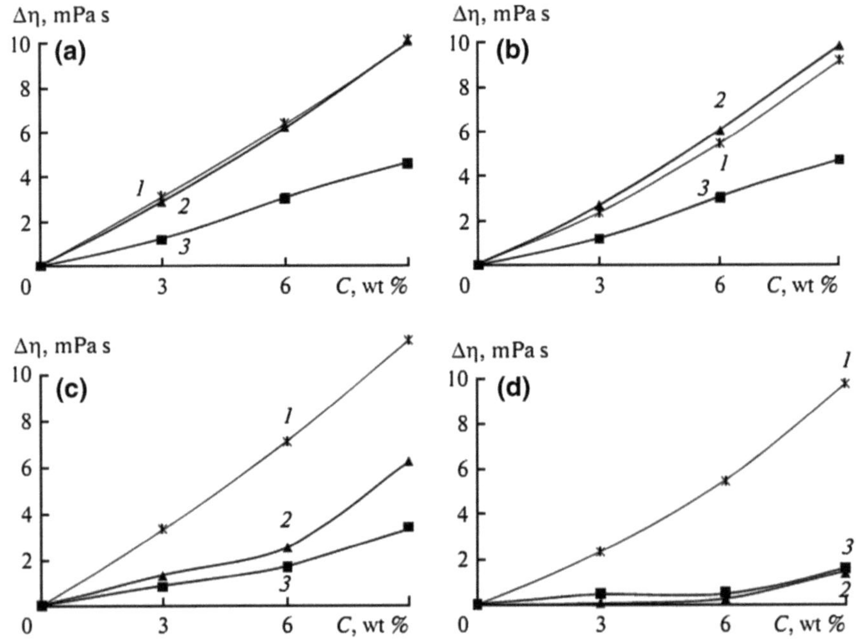

Fig. 2.17 Changes in the oil viscosity on the concentration of **a** PMMA-1, **b** PMMA-2, **c** PMMA-3, and **d** PMMA-4 modifiers with different molecular masses, introduced into the base oil PAO6 upon measuring by *1* capillary method, *2* magnetoelastic, and *3* solid-state viscometers

non-Newtonian fluids at the operating frequency (5.3 MHz) of the solid-state viscometer (at shear rate about 10^5 s^{-1}). It means that the oil PAO6 with polymer additives has a pseudoplastic behavior characterized by decreasing viscosity with increasing shear rate.

As the molecular weight of the additive increases (introduction of PMMA-3) the non-Newtonian behavior of the test oils is observed already at the operating frequency of the magnetoelastic viscometer (Fig. 2.17c). Further increase in the molecular weight of the additive (PMMA-4 VI) results in a situation when the apparent viscosities of the oils coincide at the frequencies 30 kHz and 5.3 MHz or are close to the base oil values without additives (Fig. 2.17d).

The analysis of correlations between the readings given by the magnetoelastic, capillary and solid-state acoustic viscometers for the tested base mineral and synthetic oils has indicated that the developed viscometer shows the data close to the capillary method within the range 17–500 cSt. The solid-state acoustic viscometer gives the least validity in measuring viscosity above 300 cSt. Its readings exceed considerably the anticipated data, evidently because the oil layer on its sensitive surface behaves like a gel at high operating frequency of the viscometer.

Monitoring of artificially aged mineral oil has revealed that the readings of the solid-state viscometer are much higher than viscosity measured by the magnetoelastic viscometer. This can result from the changes in chemical structure of the oil during artificial aging, and intensified chemical activity leading to the change in the interfacial interactions between the oil and sensitive surface of the solid-state viscometer. Since the detector is sensitive to the properties in a very thin layer of the oil, evidently, these interfacial interactions affect the readings.

Investigations of the effect of PMMA VI improvers on the viscometer readings have indicated that with increasing molecular weight of PMMA under the same concentration in the synthetic base, the oil starts to behave like the non-Newtonian fluid at lower frequencies of viscosity measurements. As the molecular weight of the additive increases till $M_w = 530{,}537$ Daltons, the apparent oil viscosities measured at frequencies 30 kHz and 5.3 MHz coincide and approach the viscosity values of the base oil without additives. The results obtained agree well with the conclusions presented elsewhere [35]. They have proved that mineral oils behaving as Newtonian fluids under low velocities start to behave like non-Newtonian fluids under the shear rates above 10^5–10^6 s^{-1} frequently occurring in practical applications.

Evidently, to make a correct choice of the lubricating oil with account of predicted rheological behavior in a tribosystem and its monitoring during operation, it is expedient to measure viscosity at the shear rates close to those used in the real tribosystems. In particular, it is important to estimate viscosity of non-Newtonian engine oils at medium shear rates (10^3–10^4 s^{-1}) at low temperature, and high shear rates (10^5–10^7 s^{-1}) at high temperature to make the best correlations with the engine performance [36].

2.6 Conclusions

An important aspect of ensuring the proper operation of tribosystems is the correct selection of oil, according to viscosity and its variation with temperature, as well as monitoring the oil viscosity during operation. If a change in viscosity is detected, subsequent analysis of the oil can identify the cause of such change. Either an increase or decrease in oil viscosity can result in the disturbance of the bearing capacity of tribosystem.

The review of methods of monitoring the oil viscosity shows that at present, portable viscometers requiring periodic sampling of oil are widely applied, however real-time monitoring is necessary for the reliable operation of machines. Among the devices mounted directly into the oil circulation lines and, particularly, into the lubrication systems of vehicles, acoustic devices are the most promising. Research centers carry out works for the perfection of methods for measuring the viscosity and the design of acoustic devices with the purpose of their wide application for real–time diagnostics.

In parallel with solid–state sensors based on acoustic waves, magnetoelastic viscosity sensors are currently under development; they operate at lower frequencies. Both piezoelectric and magnetoelastic types of the on-line acoustic viscometers have proved to be promising in the real-time monitoring of lubricating oils.

References

1. L.V. Markova, N.K. Myshkin, Machine tribodiagnostics (Belorusskaya Nauka, Minsk, 2005), p. 251 (in Russian)
2. L.V. Markova, N.K. Myshkin, M.S. Semenyuk, V.M. Makarenko, A.V. Kolesnikov, H. Kong, H.-G. Han, E.-S. Yoon, Methods and instruments for condition monitoring of lubricants. J. Friction Wear **24**(5), 50–59 (2003)
3. R. Serrato, M.M. Maru, L.R. Podove, Effect of lubricant viscosity grade on mechanical vibration of roller bearings. Tribol. Int. **40**(8), 1270–1275 (2007)
4. J.J. Truhan, J. Qu, P.J. Blau, The Effect of lubricating oil condition on the friction and wear of piston ring and cylinder liner materials in a reciprocating bench test. Wear **259**(7–12), 1048–1055 (2005)
5. J. Fitch, Trouble–shooting viscosity excursions. Practicing Oil Analysis Magazine Is.5 2001, http://www.machinerylubrication.com/Magazine/Issue/PracticingOilAnalysis/5/2001
6. S. George, S. Balla, V. Gautam, M. Gautam, Effect of diesel soot on lubricant oil viscosity. Tribol. Int. **40**(5), 809–818 (2007)
7. D. Troyer, Understanding absolute and kinematic viscosity, Practicing Oil Analysis Magazine Is.3, 2002, http://www.machinerylubrication.com/Magazine/Issue/PracticingOilAnalysis/3/2002
8. L.V. Markova, V.M. Makarenko, M.S. Semenyuk, A.P. Zozulya, On–line monitoring of the viscosity of lubricating oils. J. Friction Wear **31**(6), 433–442 (2010)
9. ASTM D445: Standard Test Method for Kinematic Viscosity of Transparent and Opaque Liquids (and Calculation of Dynamic Viscosity)
10. ASTM D2983: Standard Test Method for Low-Temperature Viscosity of Lubricants Measured by Brookfield Viscometer
11. J. Watson, Measuring viscosity. Hydrocarbon Engineering, Is.5, 2002, http://www.energyglobal.com/magazines/latestissue/hydrocarbon–engineering.aspx
12. D.A. Brookfield, US Patent 5503003, G01N 11/10, G01N 11/14. Portable viscometer—08/355,634. Filed 14.12.94. Published 02.04.96
13. J.R. Norcross, US Patent 5959196, G01N 11/12, G01N 11/10. In-line viscometer—08/900,170. Filed 25.07.97. Published 28.09.99
14. R.E. Kasameyer, W. Warren, US Patent 6584831, G01N 11/10, G01N 11/16. Fast-recovery viscometer—10/038,491. Filed 21.12.01/ Published. 01.06.03
15. J. Watson, Viscosity Measurement Reconsidered. Hydrocarbon Engineering Is.10, 2002, http://www.energyglobal.com/magazines/latestissue/hydrocarbon–engineering.aspx
16. H.P. Snel Dirk, Patent WO0022412, /B06B1/06, G01N11/16. Piezo-electric viscometer—WO1999NL00631. Filed 13.10.99. Published 04.20.2000
17. H.A. Wright, US Patent 6668621, G01N 11/10, G01N 11/16.Viscosity measurement by means of damped resonant vibration normal to an approximate rigid plate—10/170,071. Filed 13.06.02/ Published 30.12.03
18. L. Matsiev, J. Bennett, D.M. Pinkas, M. Spitkovsky, O. Kolosov, et al. US Patent 7043969, G01N 11/16. Machine fluid sensor and method—US 20040099050. Filed 02.06.03. Published 16.05.06
19. M. Rudkin, D. Atkinson, US Patent 4922745, G01N 9/00, G01N 11/16, G01N 11/10. Fluid transducer—07/166,254. Filed 10.03.88. Published 08.05.90
20. W.A. Farone, R.F. Sacher, C. Fleck, US Patent 6439034, G01F 1/66, G01N 11/10, G01N 11/16. Acoustic viscometer and method of determining kinematic viscosity and intrinsic viscosity by propagation of shear waves—09/699,949. Filed 30.10.00. Published 27.08.02
21. B. Drafts, Acoustic wave technology sensors. IEEE Trans. Microw. Theory Tech. **49**(4), 795–802 (2001)
22. K. Durdag, Solid state acoustic wave sensors for real–time in–line measurement of oil viscosity. Sensor Rev. **28**(1), 68–73 (2008)

23. R.P. Buck, E. Lindner, W. Kutner, G. Inzelt, Piezoelectric chemical sensors (UPAC technical report). Pure Appl. Chem. **76**(6), 1139–1160 (2004)
24. R.M. Lee, X.J. Zhang, J.M. Hammond, A remote acoustic engine oil quality sensor. Prod. IEEE Ultrason. Symp. **1**, 419–422 (1997)
25. A. Agoston, C. Otsch, B. Jakoby, Viscosity sensors for engine oil condition monitoring—application and interpretation of results. Sens. Actuators **121**(2), 327–332 (2005)
26. K. Durdag, Solid–state viscometer for oil condition monitoring. Practicing Oil Analysis Is.11, 2004, http://www.machinerylubrication.com/Magazine/Issue/PracticingOilAnalysis/11/2004
27. B. Jakoby, M.J. Vellekoop, Viscosity sensing using a love–wave device. Sens. Actuators **68** (1–3), 275–281 (1998)
28. L.V. Markova, N.K. Myshkin, H. Kong, H.G. Han, On-line acoustic viscometry in oil condition monitoring. Tribol. Int. **44**, 963–970 (2011)
29. C. Riesch, E. Reichel, A. Jachimowicz et al., A suspended plate viscosity sensor featuring in–plane vibration and piezoresistive readout. J. Micromech. Microeng. **19**(7), 2009, http://iopscience.iop.org/0960–1317/19/7/075010
30. E. Reichel, C. Riesch, B. Weiss, B. Jakoby, A vibrating membrane rheometer utilizing electromagnetic excitation. Sens. Actuators A **145–146**, 349–353 (2008)
31. J. Barandiaran, J. Gutierrez, Magnetoelastic sensors based on soft amorphous magnetic alloys. Sens. Actuators A **59**(1–3), 38–42 (1997)
32. G. Engdahl ed. Handbook of giant magnetostrictive materials (Academic Press, San Diego, 2000), p. 386
33. C. Grimes, C. Mungle, K. Zeng et al., Wireless magnetoelastic resonance sensors: a critical review. Sensors **2**, 294–313 (2002)
34. ASTM D341: Standard Test Method for Viscosity—Temperature Charts for Liquid Petroleum Products
35. G.W. Stachowiak, A.W. Batchelor, Engineering tribology. 3rd ed. (Butterworth-Heinemann, 2005), p. 832
36. D.L. Alexander, Change of high-shear rate viscosity of engine oils during use: a review. In: High-temperature, high-shear (HTHS) oil viscosity: measurement and relationship to engine operation, ASTM STP 1068, ed by Spearot JA (American Society for Testing and Materials, Philadelphia, 1989), pp. 60–73

Chapter 3
Monitoring of Water Content in Oil

3.1 Introduction

Fuel and lubricant condition monitoring in industrial equipment is one of the main means for preventing the premature wear of engines, mechanisms, and hydraulic systems. In particular, revealing of water in machine oil is one of the primary signals of the potential catastrophic wear of tribosystems [1].

Water may enter the system by different ways from the environment. Since water and mineral oil are incompatible fluids free water is accumulated at the bottom of the reservoir or in some lower places of the oil tube because of its higher mass density. During mechanism operation oil flows circulating in the system can entrap free dropped water. In the friction zone water replaces oil and disturbs the lubrication regime. In sliding bearings the damage of the hydrodynamic oil film can result in severe wear. Only 1 wt% water in oil can shorten the rated service life of a sliding bearing by almost 90% (Fig. 3.1) [2]. Water also reacts with some additives performing antiwear, antioxidative, antifoam, and washing functions in oil. Water hampers totally or partially the action of dispersants and emulsifiers that prevent the agglomeration and merging of admixtures.

The polar nature of water molecules promotes the surface corrosion of common metals, especially in the presence of air, and favors electrochemical reactions between different metals. Water affects harmfully the machine part wear but also can enhance oil oxidation that may cause its premature aging, especially in the presence of catalytic metals including copper, lead, and tin. Moreover, it is known that some synthetic oils, e.g. esters of phosphoric acid and diacids react with water resulting in the decomposition of the oil fluid base and acid formation.

Water as well accelerates the aging process for transformer oils. Water can enter the tank from outside, either being carried in as a vapor in the air or via a leaky seal. However, water is also formed inside the tank due to breakdown of the paper insulation.

Monitoring moisture content in industrial oils allows for servicing and prevention of unscheduled downtime, thus it can cut maintenance costs substantially.

N.K. Myshkin and L.V. Markova, *On-line Condition Monitoring in Industrial Lubrication and Tribology*, Applied Condition Monitoring 8,
DOI 10.1007/978-3-319-61134-1_3

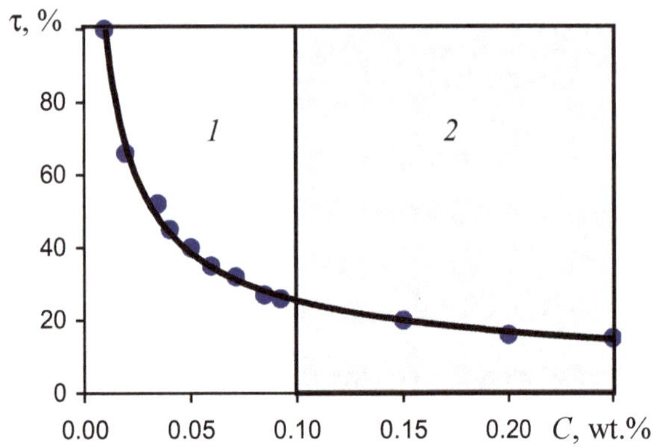

Fig. 3.1 Service life (τ) of bearing as a function of water content (*C*) in oil: *1*—transparent oil; *2*—turbid oil [2]

3.2 Forms of Water in Oil

Water can exist in oil in three forms [1–3]:

- dissolved water molecules are tied by hydrogen bonds to the hydrocarbon molecules of the oil;
- emulsified water is supersaturated in solution but has not yet totally separated from the oil (it usually gives oil a milky appearance);
- free water is also supersaturated in solution but in a high enough concentration to form water droplets and separate from the oil (see Table 3.1 based on [2] data).

Table 3.1 Forms of water in oil

Forms of water in oil		Size of dispersed particles	Optical characteristic	Notes
Dissolved water		Water hydrogen bonded to the hydrocarbon molecules of oil	Transparent	–
Emulsion	Microemulsion	0.01–0.2 μm	From transparent to semitransparent	<0.05 μm transparent ≈1 μm milky white 1–0.1 μm blue white 0.1–0.5 μm gray (semitransparent)
	Macroemulsion	0.2–50 μm	Opaque	–
Free water		Water is separated from oil	–	–

Water whose molecules are bound by hydrogen bonds to polar oil molecules and additives is considered **dissolved** water. Moreover, water dispersed in oil and formed thermodynamically stable microheterogeneous system is also refereed to as dissolved. Most industrial oils, for example hydraulic, turbine etc., can contain up to 0.02–0.06 wt% of dissolved water depending on temperature of the oil service. Aged oil is capable of retaining 3…4 times more dissolved water than the fresh one [2].

The maximal quantity of dissolved water in the oil is called the saturation level. At over-saturation water presents in oil as microscopic droplets forming an **emulsion**. In this case oil becomes turbid. There are two kinds of emulsions, i.e. macro- and microemulsions. Macroemulsions contain water droplets 0.2–50 μm in size visible in the microscope. Microemulsions contain water droplets 0.01–0.20 μm in size not visible in the microscope. At the particle diameter of 1 μm the emulsion is of milky-white color, 1.0–0.1 μm is white-blue, 0.1–0.5 μm is gray and semi-transparent, <0.05 μm the emulsion is transparent.

Further increase of water concentration in oil leads to separation of the two phases producing a layer of **free water** formed by coalescent droplets of emulsified water.

The quantity of dissolved, emulsion or free water in oil depends on the solubility defined by the water quantity that can dissolve in oil at a certain temperature.

The effect of oil aging and additives on the solubility of water in oil is shown in Fig. 3.2 [5]. With increasing concentration of polar additives in oil its water solubility rises. Therefore the introduction of additives (curve b) and oil aging (curve c) lead to increasing of the moisture adsorption compare to that of the new oil (curve a) at a given temperature.

Different oils are capable of dissolving water depending on their base and additive composition and, therefore, have different water saturation levels (Table 3.2).

Pure mineral oils are saturated with water even at the concentration $C_w = 10^{-4}$ wt%. Transformer oils contain few additives and are saturated at $C_w < 10^{-3}$ wt% of water. Hydraulic fluids are commonly saturated at C_w from 10^{-2} to 10^{-1} wt%, industrial oils—from 6×10^{-2} to 5×10^{-1} wt%, and automotive lubricants containing a wide complex of additives at C_w = 1–5 wt% [4]. The saturation of oils oxidized in operation occurs at a greater water concentration compared with fresh oils.

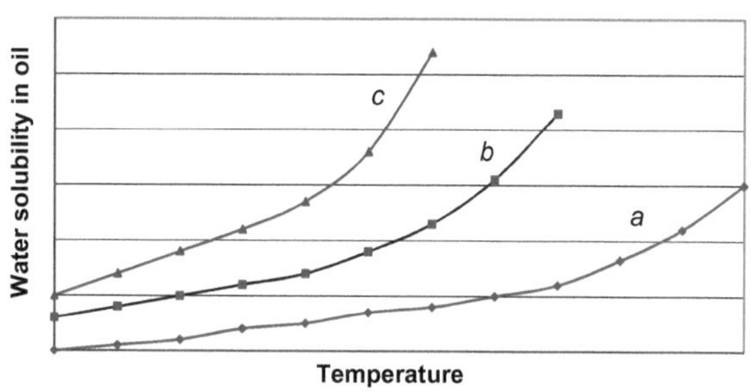

Fig. 3.2 Effect of oil aging and additives on the solubility of water in oil: *a*—base oil, *b*—base oil with additives, *c*—used oil [5]

Table 3.2 Dependence of dissolved amounts of water on lubricant type

Lubricant type	Normal water content (%)	Saturation level (%)
Automotive lubricants		1–5[b]
Diesel engine oil	0.5 (max)[a]	
Industrial oil	0.02–0.06[c]	0.06–0.5[b]
Hydraulic oil	0.05–0.15[a]	0.01–0.1[b]
Gear box oil	0.05–0.15[a]	
Turbine oil	0.05–0.15[a]	0.01[b]
Transformer oil		0.0001–0.0003[b]
Pure mineral oil		0.0001[b]

[a]www.kittiwake.com
[b]Ref. [4]
[c]Ref. [2]

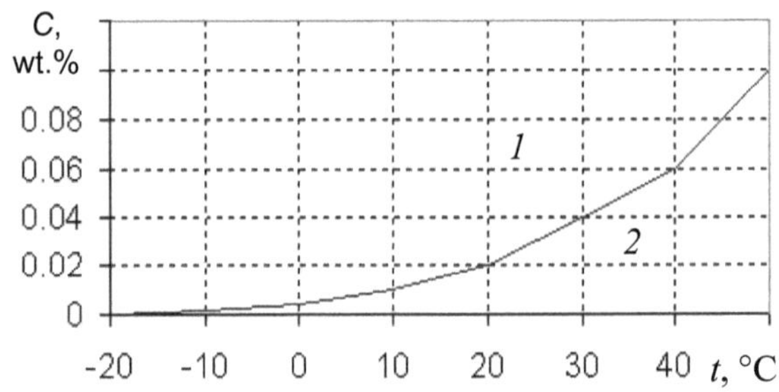

Fig. 3.3 Temperature curve of saturation of hydraulic oil with water: *1*—free and emulsified water; *2*—dissolved water [6]

The water solubility in oil increases exponentially with temperature. Figure 3.3 shows the temperature dependence of the water saturation level for commercial hydraulic mineral oil with the operation temperature of 40 °C [6]. The saturation level is 0.06 wt% at the operation temperature and 0.02 wt% at the room temperature (20 °C). Since free and emulsified water affect more harmfully compared to dissolved water it is important to prevent the appearance of free water when shut-downing the equipment (i.e. when it is cooled down to the room temperature). Therefore, in order to exclude the entering of free water when actuating the system it is needed to monitor the moment when such water concentration is achieved in the oil at the working temperature which does not exceed the saturation level at the room temperature.

3.3 Water Content Characteristics

To evaluate the water content in oil such physical characteristics are used as the mass concentration and the relative saturation.

The ***mass concentration*** of water is defined as the ratio of the water mass in the oil sample to its mass and expressed in mg/kg or in percents.

The ***relative saturation*** *RS* of oil is the ratio of the mass concentration of water in oil C_w to the saturation level C_s or the ratio of the partial pressure of water vapor *P* to the pressure of water vapor P_s in oil saturation with water at a specified temperature [3]:

$$RS = \frac{C_w}{C_s} 100\% = \frac{P}{P_s} 100\%$$

The relative saturation can be expressed in both percents and relative units. In the latter case the "water activity" *aw* is used. The values $RS = 0\%$ and $aw = 0$ means the absence of water in oil and $RS = 100\%$ and $aw = 1$ correspond to full oil saturation with water.

For example, measured water content in hydraulic oil (see Fig. 3.3) is 0.03 wt% at 40 °C. From the temperature curve we can find that water solubility is 0.06 wt% at 40 °C. Then it is possible to calculate the relative saturation by formula $RS = \frac{300}{600} 100\% = 50\%$.

Under laboratory conditions the "mass concentration" parameter, which characterizes the absolute water content in oil, is commonly used.

At the same time, since the water solubility in oil depends on its type, additives contained, operation time, and temperature, then the relative saturation is often more appropriate for evaluating the oil working characteristics. In contrast to the mass concentration the relative saturation rises with decreasing temperature (Fig. 3.4), therefore, the relative saturation is often more informative in evaluation of the oil characteristics [7].

For example, a sample of oil was taken for determination of the water content. The temperature of the oil at the time of sampling was 62 °C. The laboratory performed the analysis and determined the water content to be 11×10^{-4} wt%. From Eq. 1 it is calculated that the solubility level at 62 °C is 259×10^{-4} wt%. As discussed previously, relative saturation is the actual measured value compared to the solubility value. In this case it is 11×10^{-4} wt% divided by 259×10^{-4} wt% resulting in a relative saturation of 4.25%.

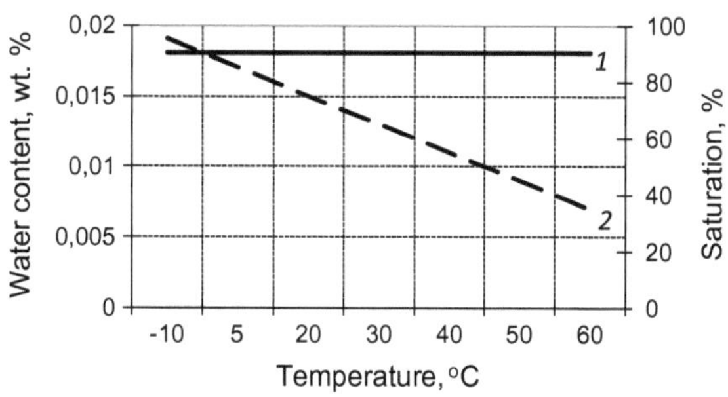

Fig. 3.4 Temperature dependence of mass water content (*curve 1*) and oil saturation (*curve 2*) with water [7]

As water saturation exceeds 90% in any system, there is a risk for segregation, especially if the temperature decreases.

3.4 Laboratory Methods for Measuring Water Content in Oil

Fourier Transform Infrared (FTIR) Spectroscopy is a common method for the laboratory determination of water content. The method is based on the fact that molecules absorb IR-radiation at characteristic wavelengths (resonance frequencies). The resonance frequencies of the molecules result from the motion of functional groups. In the water molecule (H_2O), it is the O-H functional group that contributes to the resonance frequency around 3450 cm^{-1}. However, molecules can have similar functional groups which hinders the analysis. For example, sometimes it is impossible to distinguish between the absorption of IR-radiation by water from absorption by glycol or oxidation inhibitors since the absorption bands are rather wide and located close to each other.

A list of typical resonance frequencies for common molecules measured using FTIR is given in Table 3.3 [8].

The similarity of functional groups creates a problem with FTIR. For example, if a sample is analyzed and the infrared absorption recorded in the 3600–3400 cm^{-1} region, one may not be able to differentiate between absorption due to water, glycol

Table 3.3 Common molecular species measured by FTIR [8]

Parameter	Wave number (cm^{-1})	Precision (%)	Interferences
Oxidation	Mineral oil—1750 Organic Ester—3540 Phosphate Ester—815	±25	VI improvers, dispersants Heavy moisture Contamination
Sulfation	1150	±25	Sulfurous additives (ZDDP), sulfur/phosphorus EP, some rust inhibitors
Nitration	1630	±25	VI improvers, dispersants
Soot	2000	±10	Particle size/density, engine type
Water	Mineral oil—3400 Organic Ester—3625	±25	Excessive soot, detergent additives, glycol, antioxidant, ester basestocks
Glycol	880, 3400, 1040, 1080	±50	Water, antioxidants, oxidation by-products
Fuel	Diesel—800 Gasoline—750 Jet Fuel—795–815	±50	Varying fuel aromaticity, fuel evaporation
Phenol Inhibitors	3650	±50	Moisture, glycol
ZDDP	980	±35	Aromatic impurities, fuel, silica (dirt)

contamination or antioxidant additives. Their absorptions peaks are usually fairly broad and may overlap hampering the data interpretation. Therefore, despite the sensitivity of FTIR spectroscopy to free, emulsion, and dissolved water its lower water detection limit is 0.1%. Laboratory FTIR spectrophotometers allow for analyzing oil within 600–4000 cm^{-1} frequency range. To provide early diagnostics, proper oil replacement and finding optimal maintenance terms of equipment, new laboratory devices are developed using optical methods to determine lower water contents (10^{-4} to 5×10^{-4} wt%).

Among these methods is photoacoustic spectroscopy (PAS) based on the excitation of sample molecules by light energy and the further recording of the acoustic wave generated at the moment of returning of the molecules into their basic state. PAS has a considerably lower water detection limit than other purely absorption kinds of spectroscopy.

A diagram of the PAS experimental arrangement is presented in Fig. 3.5 [9]. Oil flow passes through the layered-prism cell with a sapphire entrance prism, quartz exit prism, and an optical pathlength of 0.76 mm. Acoustic contact between the transducer (6 mm diameter, 5 MHz frequency, Panametrics) is facilitated by an ultrasonic couplant fluid (Sonic Instruments). Excitation light of 2.93 μm (3415 cm^{-1}) is generated by Raman shifting (900 psi deuterium in 1 m cell from a pulsed Nd-YAG laser (λ = 1064 nm) operating at 20 Hz with 8 ns pulse width. Because the Raman conversion generates several anti-Stokes (807, 650, 544 nm, etc.) and Stokes (1561, 2928 nm, etc.) lines, optical filters and wavelength-selective mirror are used to remove or reflect the unwanted Raman lines. The signals from the transducer are amplified using the preamplifier (Panametrics, model 5670, 40 dB) and the waveforms collected on the digital oscilloscope (Lecroy, model 9362). The photoacoustic signal is determined by either calculating the maximum peak-to-peak voltage of the waveform or by integrating the signal voltage over the time period selected by the boxcar. The photoacoustic cell is flushed with methylene chloride and dried with N_2 gas between the samplings.

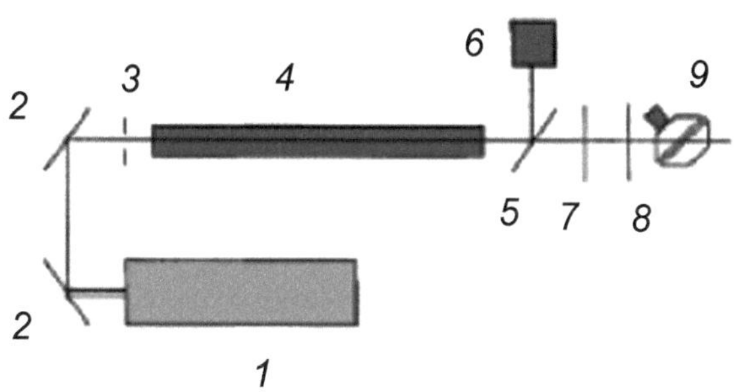

Fig. 3.5 Schematic of photoacoustic experimental setup: *1*—Nd-YAG laser at λ = 1064 nm, *2*—1064 nm high power laser mirror, *3*—iris diaphragm, *4*—1 m Raman cell filled with deuterium at 900 psi, *5*—coated-sapphire 1064 nm selective mirror, *6*—1064 nm beam dump, *7*—optical filter passing light >845 nm, *8*—optical filter passing 2670–8000 nm light, *9*—layered-prism cell with input prism of sapphire, output prism of quartz, and 5 MHz contact transducer [9]

The PAS method allows measuring the water content to 0.0045, 0.006, and 0.0515 wt% in transmission, hydraulic, and synthetic oils, respectively.

The laboratory optical methods such as FTIR spectroscopy and photoacoustic spectroscopy are quite complex and expensive. As a rule, FTIR spectroscopy is used for qualitative assessment and is supplemented by other water checking analyzes, e.g. the Karl Fischer method.

The Karl Fischer Method (titrimetry) is the most precise and widely used to determine the content of free, emulsion, and dissolved water in lubricants. The method is based on measuring the reagent quantity necessary for reaction with a certain component of the solution corresponding to the stoichiometry of chemical reactions between them. In analysis the volume V of the solution being analyzed is checked and either the volume or the mass of the titrant is added, i.e. the solution with the exactly known concentration C_t of the reagent. During titration, the equilibrium concentrations of the reagent, substance to be analyzed, and reaction products change. By measuring the titrant volume V_t at the end of titration the concentration C_w of the solution is calculated by the formula $C_wV = C_tV_t$. The final point of titration is found by variation in some physical characteristic of the solution depending on the concentration ofthe substance to be determined.

There are two Karl Fischer titration methods, i.e. volumetric [10] and coulometric [11]. The main difference is that in the former method the titrant is added to an oil sample while in coulometric method the titrant is formed electrochemically in the titration cell. The final point of titration is found by the oxidation-reduction potential (potentiometric titration) in the volumetric method or by the quantity of electricity in the coulometric method. The water content is calculated by the quantity of the reagent selectively reacting with water and spent on the involvement of the whole volume of water in the reaction. The volumetric titration is commonly applied in oil analysis. It uses potentiometric cell to register the end moment of the reaction and assess the reagent quantity. The method is capable of detecting relatively high mass concentration of water, above 0.03 wt%.

Recently the coulometric method that measures low water concentrations draws more attention. However, the known Karl Fisher reagents allow for analyzing only additive-free oils. To analyze oils containing additives only the distillation technique is appropriate which provides water evaporation from oil during its vacuum heating, condensation and dissolution of water in toluene followed by titration. Such a technique makes possible determining the water concentration of no less than 10^{-3} wt%.

The method of measuring the water content using reactions of a sample of the wet material with different **calcium compounds** is more easy-to-implement. The mass of the reaction product, e.g. acid or gas is proportional to the water mass content in the sample. The device comprises a container into which the oil sample and calcium hydride are placed. After intensive mixing the quantity of the gas released in the reaction is registered by the pressure gage. The method is suitable for analyzing motor, hydraulic, and transmission oils and is capable of determining the water content within 0–10 wt% range.

The **visual crackle test** is the simplest method. It consists in the deposition of the oil drop to the plate heated to 160 °C. The number of bubbles arising in oil when the water is evaporated indicates water presence. Small number of bubbles indicates the water content of 0.05–0.1%. A greater number of larger bubbles can point to ≈0.1–0.2 wt% water content; at $C_w > 0.2\%$ crepitating sounds are heard.

Thus, nowadays the commonly employed laboratory techniques for water content in oil evaluation are the following:

- FTIR Spectroscopy used for screen test, measurement range is higher then 0.1 wt%;
- photoacoustic spectroscopy is quite complex and expensive, allows to determine lower water contents (10^{-4} to 5×10^{-4} wt%).
- Karl Fisher technique is rather complex and permits to measure water content above 0.03 wt% by volumetric method and lower 0.01 wt% by coulometric method;
- calcium hydride test—rather simple and low cost technique with lower limit 0.005 wt%, although the running costs must be considered with regard to the reagents, as well as the health and safety issues of these reagents
- visual test—the simplest technique permitting to carry qualitative estimation of water content higher then 0.05 wt%.

The main disadvantage of the laboratory methods is that the results are obtained with some delay and sample preparation is needed. Often the real-time monitoring of the water content in oil is more reliable.

3.5 Methods and Devices for Real-Time Monitoring of Water Content in Oil

Real-time monitoring devices/sensors can be divided into two groups, i.e. those based on the analysis of variations in oil characteristics (electrical or optical) and those in which sensors include the elements absorbing moisture from the test oil and further analysis of variations in the element characteristic is followed (Fig. 3.6).

Devices without the absorbing element

Devices which do not extract water from the test oil employ electrical or optical properties of water. Fresh mineral oil consists of mainly hydrocarbon molecules with the weakly polar or non-polar distribution of charges which causes its high electric resistance and relatively low permittivity. The permittivity of mineral oils is 2.0–2.3, synthetic oil ≈6.9, and water ≈75. The relation between the water content and permittivity is used in the devices with capacity sensors. The optical properties dependence on water content, particularly, variation in absorption of IR radiation at specific wave lengths, are used in water sensors built in oil circulation line.

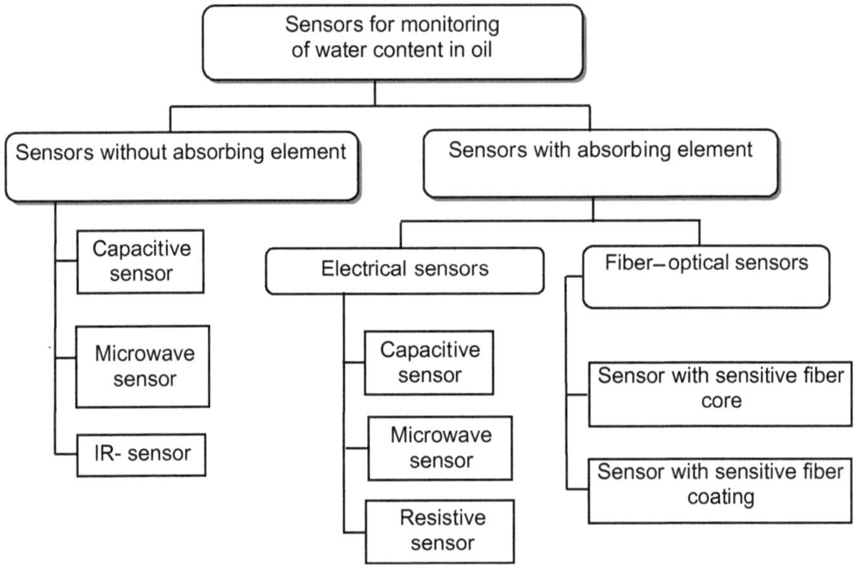

Fig. 3.6 Classification of built-in devices of water content in oil

Devices with capacitive sensors. The capacitive sensor is a measuring capacitor in which oil acts as a dielectric. Variation in the water content in oil causes changes in capacitance and, respectively, characteristics of the measuring circuit, i.e. the oscillator frequency, circuit current, etc. used as informative parameters.

Methods and apparatuses based on measurement of *variations of oscillator frequency* are described elsewhere [4, 12].

Sensor of the device is formed by plate 1 with three elongated electrodes 2 arranged in a symmetrical pattern at the points of an equilateral triangle (Fig. 3.7a) [12]. Capacitors formed by the electrodes are parts of the ring oscillator circuit. The electrodes are immersed in oil. Water increases the oil permittivity and, respectively, varies the capacity and oscillator frequency. The frequency is decreased by means of frequency divider and then it is converted into the output voltage which is compared to predetermined limit by comparator generating the alarm signal when voltage signal exceeds the limit (Fig. 3.7b).

For marine applications, the system has been calibrated so that the signalization is actuated at 10 wt% water content in oil. The sensor system is highly effective for warning the operator to shut down the engine before catastrophic gear box damage occurs.

The measurement of electrical current at charge/discharge of the capacitor containing test oil as dielectric medium is applied in the other device for water content estimation [13].

A scheme of the device is shown in Fig. 3.8. It includes variable voltage source 2, the capacitor 1 with the test oil, resistor 5, analog to digital converter ADC, and microprocessor and temperature sensor T. Measuring capacitor 1 is modeled as the

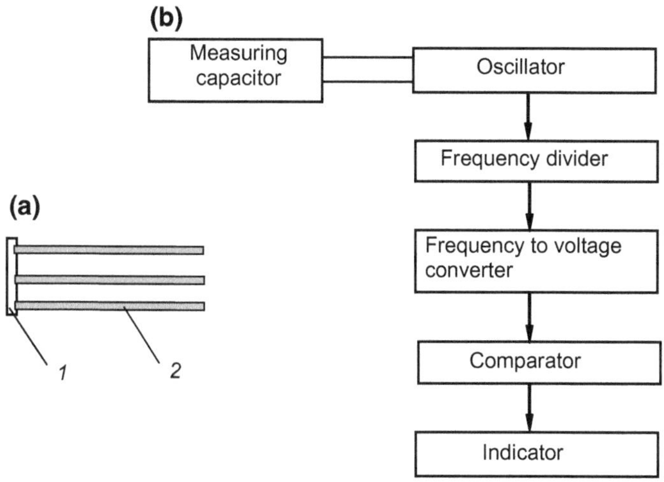

Fig. 3.7 Water content sensor based on measurement of variations of oscillator frequency: **a** measuring capacitor: *1*—plate, *2*—electrodes, **b** signal processing scheme [12]

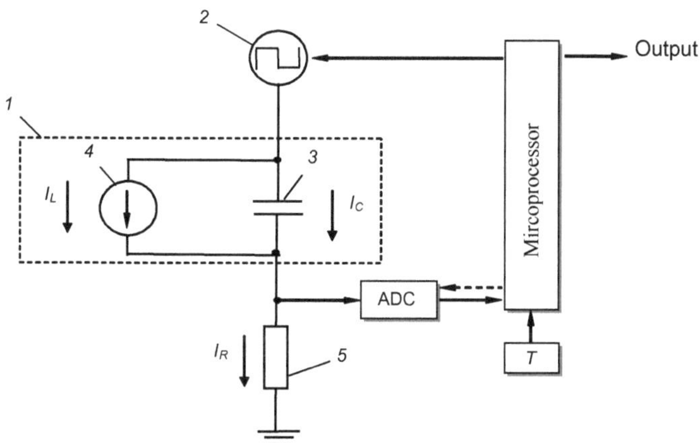

Fig. 3.8 Capacitive sensor measuring parameters of oscillation circuit at charge/discharge of the capacitor: *1*—measuring capacitor, *2*—variable voltage source, *3*—capacitor with test oil, *4*—constant current source, *5*—resistor, *T*—temperature sensor [13]

reference capacitor 3 with the test oil having a dielectric permeability of ε and constant current source 4 parallel to the capacitor. The source current has the leakage value I_L. Variable voltage source 2 induces the reactive I_C and leakage I_L currents flowing through the capacitor 1 and value of the reactive current I_C depends on oil dielectric permeability. Resistor current I_R flowing through resistor 5 is the sum of the reactive current I_C and the leakage current I_L.

Voltage produced across the resistor 5 is feed through ADC to microprocessor for further processing. Microprocessor controls the switching of variable voltage source after transition time.

The signal processing technique is the following. Current value $I_R(t)$ flowing through a capacitor undergoes an exponential decay as time increases:

$$I_R(t) = I_p e^{-\frac{t}{RC}} + I_L$$

where I_p is the peak value, R is the resistance of the series resistor, and C is the capacitance of the capacitor.

A rate of decay is determined by a time constant τ that is a product of capacitance and resistance of the circuit $\tau = RC$. Value of the leakage current I_L is defined as $I_L = I_R(t)$ at time $t \approx 5\tau$ when $I_L = I_R(t)$ current $I_R(t)$ would approach a constant value.

Reading three values of current I_p, I_L and $I_R(t)$ at known time $t = t'$ it is possible to calculate the capacitance C as follows:

$$C = \frac{t}{R \ln\left(\frac{I_R(t) - I_L}{I_p}\right)}.$$

Once the capacitance has been calculated, the dielectric permittivity of the fluid can be calculated upon the capacitance and geometry of the capacitor sensor.

Devices with microwave sensors. Capacitive sensors are rather sensitive to the contamination of the capacitor plates that reduces the measurement reliability. The disadvantage is overcome in the devices with microwave sensors operating at high frequencies from hundred megahertz to tens gigahertz. Their basic informative parameter is the amplitude or power of the wave passed through the layer of the oil under monitoring or reflected from its surface. The energy absorption by oil depends on the water content in it.

Devices with microwave sensors have different designs of volume resonators or emitters. Diagram of one of them is presented in Fig. 3.9 [14]. Microwave sensor of

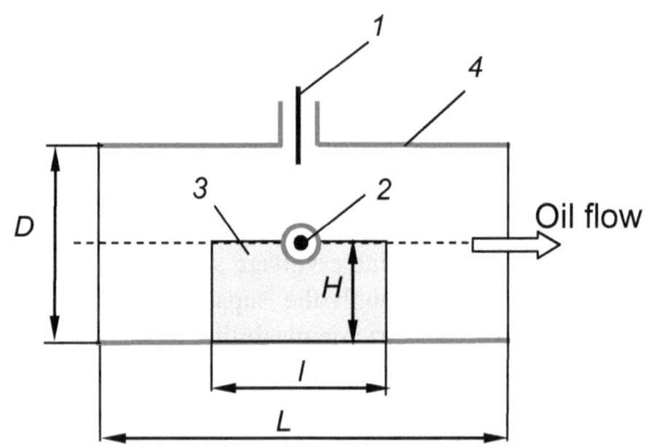

Fig. 3.9 Microwave sensor: *1* and *2*—emitter and receiver of microwave radiation; *3*—conducting plate; *4*—tube [14]

the device is the resonator including the tube 4 of length L and diameter D, transmitter 2 and receiver 1 of microwave radiation, and the conducting plate 3 of height H and length l installed in the tube 4. The dimensions of the tube 4 and the plate 3 are calculated taking in account the parameters of electromagnetic microwave to form a standing wave. The test oil flows through the tube 4. When water content in oil increases the dielectric permeability increases and as a result the resonance frequency of the sensor varies. Concentration of water in oil is estimated by shift of the measured resonance frequency using a calibrating plot. This device is rather chip but its accuracy is low, ±2 wt%.

The disadvantage of the devices with capacity and microwave sensors is that they assess the integral change of the oil permittivity caused by numerous factors that makes it difficult to select the water content contribution.

The device with an IR-sensor. The device with an IR-sensor based on the monitoring of variation in oil optical characteristics is built in the oil line and, in contrast to the FTIR spectrometer, employs a narrow spectral range [15]. An increase in the water concentration in oil raises the intensity of radiation absorption at the IR-wavelength selected from the water absorption band. The device comprises measuring and reference channels. The absorption of IR-radiation is determined by the difference of the output signals of these channels. The mass concentration of water in oil is assessed by the absorption intensity using the calibration curve. The device allows for monitoring the water content up to 2 wt% in lubricating and hydraulic oils in real time.

Devices with the Absorbing Elements

The monitoring devices without absorbing elements are distinguished by expeditious analysis; however they are capable of measuring only relatively high water concentrations. To increase the sensitivity of the operative monitoring devices the method is used that employs water accumulation in an absorbing element up to the equilibrium state that depends on the water content in oil. The amount of water accumulated in the absorbing element is measured by different electrical and fiber-optical devices (Fig. 3.6). These instruments are currently most widely used.

Electrical devices

Electrical devices are commonly based on capacitive, microwave, and resistive sensors.

Device with capacitive sensor. The capacitive sensor of the water content consists of the sensitive dielectric layer *4* absorbing water from oil and placed between the galvanically isolated conducting layers *1* and *2* (Fig. 3.10) [16]. Layer *2* has arbitrarily located grooves *3* through which oil can penetrate into the dielectric layer. This multi-layer system forms the measuring capacitor and it is fabricated on the glass substrate *5* with the sensitive layer made of thin porous polymer film. When water from oil is absorbed by the film its permittivity varies

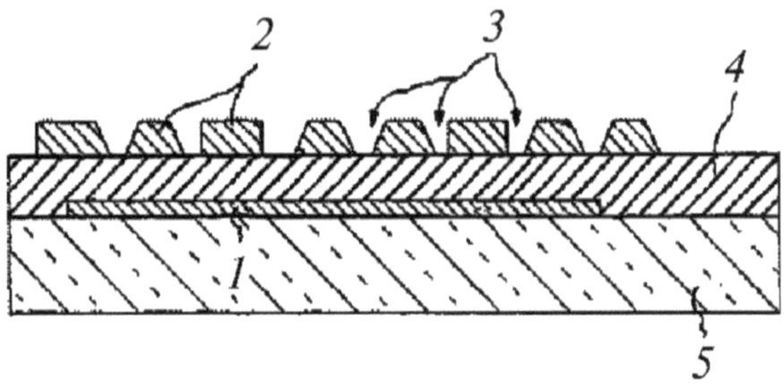

Fig. 3.10 Water capacitive sensor: *1* and *2*—conducting layers; *3*—grooves; *4*—absorbing layer; *5*—substrate [16]

that changes the capacitance and, respectively, the output signal in the electric circuit comprising the capacitor. The sensor of this type is widely used in devices produced by Vaisala (HMP228), Pall (PallWS03), DOMINO Companies, etc. The devices measure the relative saturation of oil with water in oil circulation systems and in transformer oils.

Device with the microwave sensor. To analyze variations in the properties of the sensitive layer when absorbing water from oil the microwave sensor can be applied. Figure 3.11 shows a planar immersed microwave sensor comprising the material *3* absorbing water, heater *2,* and transmission line *1* [17]. Aluminum oxide is used as the absorbing material which has the sufficient mechanical strength despite its porosity. The heater *2* installed in the porous material causes desorption of the absorbed water into the oil thus providing the periodic cleansing of the absorbing material. The sensor is the load of the transmission line. Water absorbed by the porous material varies its permittivity that causes changes in the transmission of the microwave energy and amplitude of the output signal. Water content in the test oil is estimated using the preliminary received calibrating data.

Device with the resistive sensor

Operation of such devices is based on the effect of water content on conductivity of sensitive layer which is registered during the absorption of water contained in

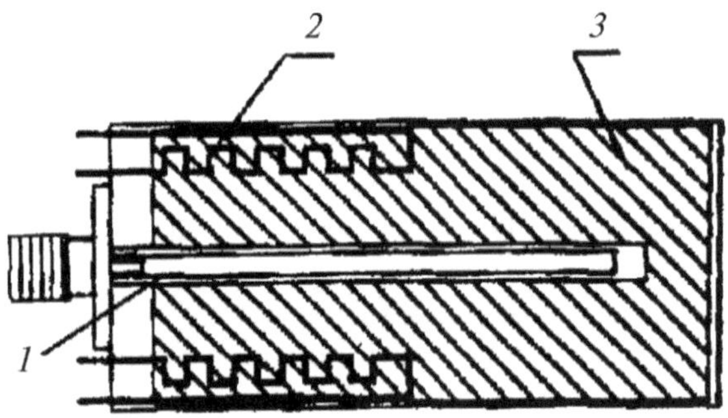

Fig. 3.11 Microwave sensor: *1*—transmission line; *2*—heater; *3*—layer of selectively absorbing material [17]

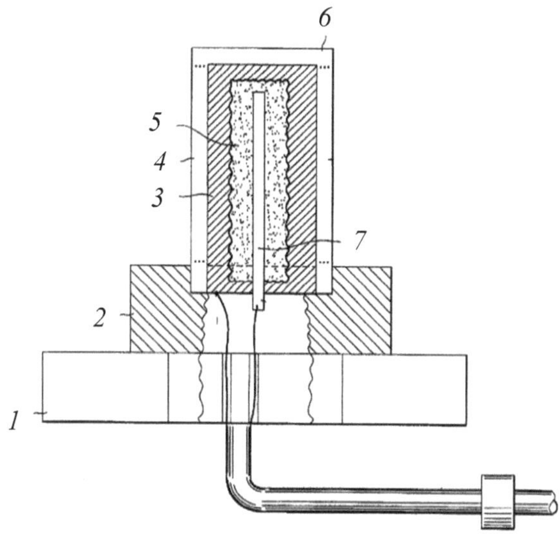

Fig. 3.12 Resistive sensor witτh polymer bead matrix: *1*—drain plug; *2*—screw; *3*—mesh; *4*—holes for oil flow; *5*—polymer bead matrix; *6*—housing; *7*—metallic rod [18]

oil [18]. The layer is a matrix of polymer bead placed between the conductive surfaces. Figure 3.12 shows the sensor that is mounted with the screw *2* to drain plug *1* in the engine oil pan. The plastic housing *6* comprises the wire mesh *3* that holds the polymer bead matrix *5*. The oil flow passes the holes *4* through the mesh *3* and polymer beads *5*. The metal rod *7* acts as the first electrode and mesh *3* as the second electrode of the electric circuit measuring conductivity of the polymer matrix. The matrix consists of polyester balls 40–70 μm in diameter cross-linked by 8% divinyl benzene and impregnated with charged groups, i.e. sodium (cation group) and sulfite (anion group). To measure water content, the one-layer matrix of beads is applied, bead swelling and shrinkage being used in the sensor. The initial swelling of the balls is produced before loading them into the sensor by immersing in a highly polar solution. As the sensor is immersed in oil, the bead size and, respectively, the conductivity (resistance *R*) of the sensitive layer vary ($R_1 < R_0$) with the water content in oil (Fig. 3.13). The sensor Oil Sentinel produced by Voelker Sensors Inc. [19] is capable of registering the water content above C_w = 0.06 vol.% (Fig. 3.14). The method is efficient at the oil temperature from 20 to 70 °C.

Fiber-Optical Devices. The passage of light through a fiber is governed by the ratio of the refractive indices of the core, cladding, and medium around the cladding. Fiber-optical devices use the gages with the sensitive fiber core or cladding.

The device with the sensitive fiber core (Fig. 3.15) includes source *1* and detector *6* of optical radiation connected by a fiber-optical line [20]. The latter consists of the transmitting *2*, detecting *5*, and absorbing *3* fibers jointed by the optical socket *4*. The absorbing fiber takes part of the fiber core and oil under

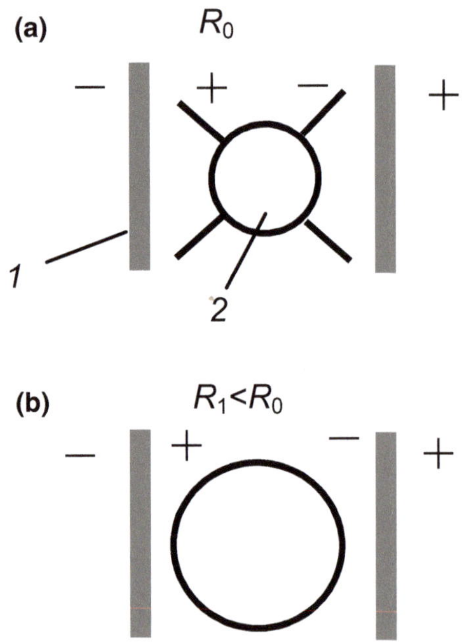

Fig. 3.13 Behavior of polymer matrix in water-free oil (**a**) and in oil with increasing water content (**b**): *1*—electrodes; *2*—polymer bead [18]

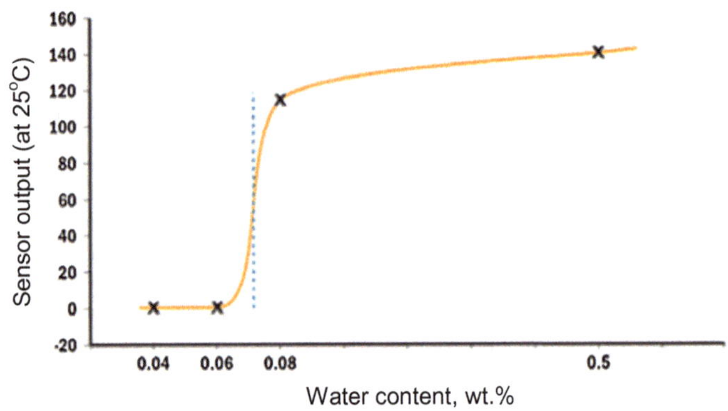

Fig. 3.14 Water content measurements by the device with resistive sensor [19]

analysis in which the fiber is immersed takes part of the coating (reflecting medium). Fiber *3* is made of an absorbing polymer or hydrogel containing a hydrochrome moisture-sensitive material. Water diffuses in the absorbing polymer until the equilibrium state occurs varying the optical properties of the polymer. The

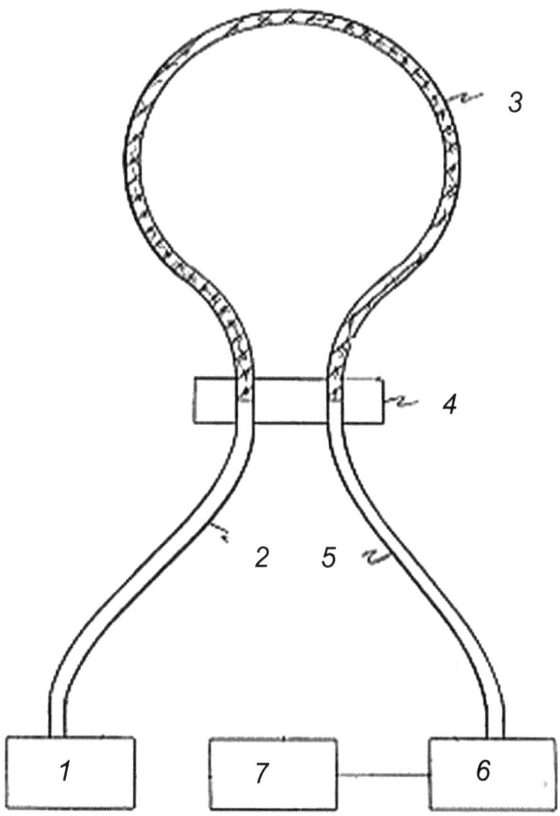

Fig. 3.15 Designs of fiber optic sensors with sensitive core: and *6*—source and detector of optical radiation, respectively; *2* and *5*—transmitting and detecting fibers, respectively; *3*—sensitive absorbing fiber; *4*—optical socket; *7*—indicator [20]

intensity of the light passed through the absorbing fiber is measured by the photodetector *6* and the output signal is transmitted to the indicator *7*.

The fiber optic devices have been also developed with gage fibers having the moisture-sensitive coatings [21, 22]. The device is based on employing variation in the refractive index and, respectively, the intensity of passed radiation when water is absorbed by the coating material from the oil.

The application of the fiber-optical devices is advisable in equipment with a high level of electrical noise, e.g. in transformers and generators. Their disadvantage is high labor-consumption and absorbing fiber manufacturing costs.

Device Based on Water Evaporation. In addition to the above electrical and fiber optic instruments with absorbing elements we can also mention a device in which water contained in oil is accumulated in air above the oil surface.

Equipment monitoring the water content in oil and holding it at the specified level is shown in Fig. 3.16 [23]. The equipment includes the device 5 based on

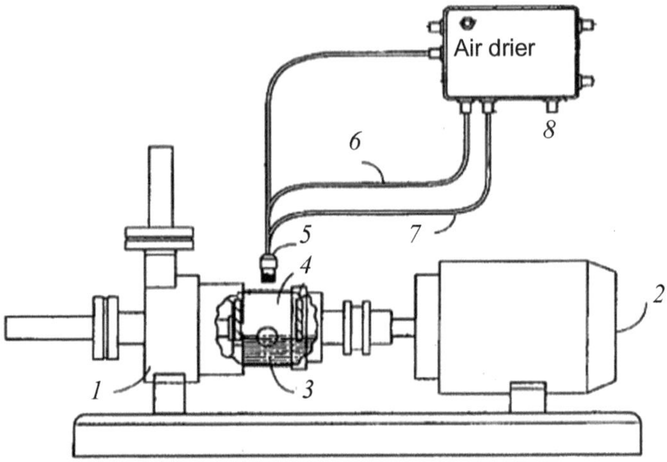

Fig. 3.16 Equipment with the in-line moisture device based on evaporation of water from oil: *1*—centrifugal pump, *2*—engine, *3*—oil sump, *4*—free volume above oil surface, *5*—moisture device based on evaporation of water from oil, *6*—line supplying the dry air, *7*—line discharging the moist air [23]

water evaporation. The device is installed in the oil sump 3, where the oil is mixed by the centrifugal pump 1. Dried air from the air drier module is periodically supplied into the vessel 4 above the oil surface through the line 6. As water is evaporated from the oil being mixed, the air humidity increases until the equilibrium state occurs when the pressures of water vapors in air and oil are the same. Using an air humidity sensor mounted in device 5 the water content in oil is determined. Measured value of water content in oil is compared with critical level. When water content increased up to critical level the humid air from the volume 4 is discharged to the air drier module through the line 7. The extracted water is removed from the air dryer module through a drain port 8 and dried air is supplied into the vessel 4 again by the line 6.

The technique could be used for monitoring of mineral and synthetic oil condition. An important feature of the device is that it carries out predictive maintenance while controlling the water content in the oil at a given level.

Commercial water monitoring devices

The main built-in water sensors, which are available in the market currently, are presented in Table 3.4.

Table 3.4 Built-in commercial devices

Device	Measuring principle	Water content	Application
HMP228 Vaisala, Finland www.vaisala.com	Capacitive measurement through an absorbing polymer film	0…1 aw 0 … 80×10^{-4} wt%	Lubrication oil Transformer oil
PallWS03 Pall Water Sensor www.pall.com	Capacitive measurement through an absorbing polymer film	0…100% RS	Petroleum and synthetic fluids and fuels
DOMINO www.double.com	Capacitive measurement through an absorbing polymer film	0…100% RS	Transformer oil
LiqMS21231 www.fullsense.com China	Capacitive measurement through an absorbing polymer film	0…1 aw 0…80 to 550×10^{-4} wt%	Transformer oil Machine oil
HYDAC AS 2000 series AquaSensor www.hydacusa.com	Capacitive measurement through an absorbing polymer film	0…100% RS	Hydraulic oil Lubrication oil
WGM 1 www.eppensteiner.com	Capacitive measurement through an absorbing polymer film	0…1 aw	Hydraulic oil Lubrication oil
WSPS 02 www.aeton.com	Capacitive measurement through an absorbing polymer film	0…100% RS	Hydraulic oil Lubrication oil
Oil Sentinel Voelker Sensors Inc. www.vsi-oil.com	Polymeric bead matrix containing charged ion groups is held between two conducting permeable surfaces	>0.08 vol.%	Industrial oils (Hydraulic oil, spindle oil)
OWD Water Cut Meter Kam Controls Inc. www.kam.com	Microwave sensor	0…100% RS With 1% of full scale accuracy	Petroleum fluids

3.6 Conclusions

Water may enter the system from the environment and oil flows circulating in the system can entrap free dropped water. In the friction zone water replaces oil and disturbs the lubrication regime, so the presence of water in machine oil is one of the reasons for severe wear of tribosystem. Monitoring of the water content in industrial oils allows for servicing and prevention of failures of equipment.

The most common devices for the on-line real time monitoring of the water content in oils are nowadays based on capacitive sensors and polymer absorbing films. Their main disadvantage is poor measurement reliability when the oil saturation with water exceeds 70%. However, their advantage is in cost-efficiency and simple maintenance. The devices with absorbing elements determine the relative oil saturation with water. Yet, if it is required to evaluate the absolute water content the devices without absorbing elements should be used.

The development of the reliable sensors installed in the equipment is of urgency. The problem consists in producing cheap sensitive sensors for detecting water in oil and providing the high-effective monitoring of the lubrication regime of friction units. Another important task is to provide the predictive maintenance with controlling the level of ware content at a given level providing the long life and reliable operation of the equipment.

References

1. L.V. Markova, N.K. Myshkin, V.M. Makarenko, M.S. Semenyuk, H. Kong, H.-G. Han, E.-S. Yoon, Monitoring of water content in oil as a method of tribodiagnostics. J. Friction Wear **25** (6), 35–43 (2004)
2. M. Barnes, Water—the Forgotten Contaminant. http://www.maintenanceworld.com/water-the-forgotten-contaminant
3. L. Lewand, Understanding water in transformer systems, the relationship between relative saturation and parts per million (ppm). Available: http://www.netaworld.org/sites/default/files/public/neta-journals/NWsp02%20-%20ChemPersp.pdf
4. R. Garvey, A. Carey, US Patent 5656767, G01N 27/22, G01N 33/26, G01N 33/28. Automatic determination of moisture content and lubricant type—No. 08/614,817. Filed 08.03.96. Published 12.08.97
5. Measuring moisture in oil. Available: http://www.vaisala.com/Vaisala%20Documents/Application%20notes/MeasuringMoistureinOil-FAQs-B210963EN-A.pdf
6. Online sensor measures dissolved water in lubricating oils and hydraulic fluids. Available: http://www.machinerylubrication.com/Read/434/dissolved-water-oil-sensor
7. Are you sensing moisture problems? Pall's new water sensor reports dissolved water concentrations. Pract. Oil Anal. **7** (1999). Available: http://www.machinerylubrication.com/Read/50/pall-water-sensor
8. Fourier transform infrared spectroscopy, Noria Corporation. Pract Oil Anal. **3** (2002). Available: http://www.machinerylubrication.com/Read/305/fourier-transform-infrared-spectroscopy
9. N.S. Foster, J.E. Amonette, T. Autrey, J.T. Ho, Detection of trace levels of water in oil by photoacoustic spectroscopy. Sens. Actuators, B **77**, 620–624 (2001)
10. ASTM D1744: Standard Test Method for Determination of Water in Liquid Petroleum Products by Karl Fischer Reagent
11. ASTM D6304: Standard Test Method for Determination of Water in Petroleum Products, Lubricating Oils, and Additives by Coulometric Karl Fischer Titration
12. A.F. Santa Maria, P.K. Banerjee, J. Vincent, US Patent 5642098, B60R 16/02, B60Q 001/00. Capacitive oil water emulsion sensor system—No. 08/634,272. Filed 18.04.96. Published 24.06.97

13. G.A. Marszalek, US Patent 6509745, G01R 31/0, G01R 027/26. Method and apparatus for measuring liquid dielectric behavior—No. 09/669,162. Filed 25.09.00. Published 21.01.03
14. E. Nyfors, A. Bringsvor, US Patent 6466035, G01N 22/00, G01R 027/32. Microwave fluid sensor and a method for using same—No. 09/701,185. Filed 04.01.01. Published 15.10.02
15. OILAN A4 water in oil analyzer. Available: http://www.kytola.com/products/oil-analyzers/oilan-water-in-oil-analyzer
16. L. Stormbom, US Patent 5075816, G01N 27/22, H01G 007/00. Capacitive humidity sensor construction and method for manufacturing the sensor—No. 07/553,596. Filed 18.07.90. Published 24.12.91
17. B.N. Scott, US Patent 6630833, G01R 1/067, G01R 29/08, G01N 22/00, G01R 027/04. Measurement by concentration of a material within a structure—No. 09/803,716. Filed 09.03.01. Published 07.10.03
18. P.J. Voelker, J.D. Hedges, US Patent 5789665, G01N 33/28, G01N 33/26, G01N 027/26. Oil quality sensor for use in a motor—No. 08/637,878. Filed 25.04.96. Published 04.08.98
19. Oil Sentinel—Theory of Operation. Water Contamination and Oxidation Measurements in Base Oils & Hydraulic Fluids, Industrial Series Sensor System, Voelker Sensors. Inc. Available: http://www.vsi-oil.com/Sentinel/OilSentTheoryOper.PDF
20. P.J. Hood, C.M. Theodore, A.M. Yates, US Patent 6582658, G01N 21/77; G01N 21/81, G01N 021/00. Fiber optic moisture sensor—No. 09/598,738. Filed 21.06.00. Published 24.06.03
21. S.H. Peterson, A.R. Pebler, R.K. Sadhir, H. Pearce Jr., C.C. Claiborne, US Patent 4894532, G01N 21/77; G01N 21/35, H01J 005/16. Optical fiber sensor with light absorbing moisture-sensitive coating—No. 07/173,897. Filed 28.03.88. Published 16.01.90
22. J.H. Rodriguez-Rodriguez, F. Martinez-Pinon, J. A. Alvarez-Chavez, D. Jaramillo-Vigueras, Polymer optical fiber moisture sensor based on evanescent-wave scattering to measure humidity in oil-paper insulation in electrical apparatus, in *Proceedings of IEEE Sensors* (2008), pp. 1052–1055
23. B. Rake, US Patent 6447573, F16N 29/00, B01D 053/02. Apparatus and method for lubricant condition control and monitoring—No. 09/670,852. Filed 27.09.00. Published 10.09.02

Chapter 4
Control of Soot Concentration in Oil

4.1 Introduction

In the course of engine operation the lubrication oil is contaminated, particularly, by products of incomplete combustion (soot). These products reach the engine oil from the combustion chamber during operation. While the presence of soot is normal, its concentration or dispersity can be abnormal, signaling an engine problem and need for the oil change.

The effects caused by the carbon soot contamination of engine oils have been investigated by many scholars [1–8]. George et al. [1] used statistically designed experiments to study the effect of soot contamination on the engine oil viscosity. Their results have shown that viscosity increases nonlinearly with soot concentration at 40 °C, but linearly with soot concentration at 90 °C. A similar increase in oil viscosity due to soot concentration has been obtained by Kornbrekke et al. [2], Agoston et al. [3], and Liu et al. [4]. Gautam et al. [5] have investigated the effect of soot-contaminated oil on wear by factorial experiments. Their statistical analyses have shown that wear increases with higher soot concentration and decreases with higher phosphorus levels of additives. In another study, Kaneta et al. [6] have shown that soot acts as an abrasive in wear of elastohydrodynamic tribosystems. Fitch [7] highlighted some engine operation problems related to soot as (a) the combustion efficiency drop with soot generation rate and (b) the oil viscosity increase with soot load and risks to engine reliability. Also, Aldajah et al. [8] in a study of diesel engine oils from an exhaust gas recirculation system observed that, despite lower friction, the test ball wear in used oil was two orders of magnitude larger than wear in clean oil.

Therefore, it is evident that soot in oil is an important parameter in the condition monitoring of diesel engines.

N.K. Myshkin and L.V. Markova, *On-line Condition Monitoring in Industrial Lubrication and Tribology*, Applied Condition Monitoring 8,
DOI 10.1007/978-3-319-61134-1_4

4.2 Up-to-Date Requirements to Monitoring of Diesel Oil

Up-to-date small-size diesels are used more widely in European cars owing to their high efficiency in fuel consumption and low fuel cost compared to those of gasoline engines. At the same time environmental-protection organizations impose more severe demands to the standards of exhausts of nitrogen and carbon oxides both of car engines and heavy-duty diesel engines used in trucks and buses. Thus, it is expected that new standards will reduce contamination of the ozone layer by 1.1 million tons per year, decrease acid rains and solid particle content in air by 43 thousand tons per year [9].

The situation has given rise to research and works intended to minimize exhausts and noise of diesel engines. A great variety of types, designs, fields of application, and operation conditions of diesel engines cause the development of the wide range of diesel oils. The type of the diesel, its boost, severity of working conditions, and quality of the fuel are the most important criteria governing the oil selection. The recommendations of oils and service schedule including the terms of oil replacement are given in service instructions of diesels. The diesel oils are based on distillate, residual or compounded base oils of selective refining obtained from sweet or sulfur-containing petroleum.

The specificity of oils results from the operation cycle of diesel engine which differs essentially from the process used in gasoline engines. In diesels mixing occurs hundreds times faster than in gasoline engines, so the mixture quality being worse. As a result, it does not burn completely and in exhausted gases there is a lot of soot, so it enters the oil, thus accelerating its aging. The soot particles are spread in the engine by the oil, they choke filters and are deposited on engine parts. To decrease the contamination it is necessary to keep the particles in the suspended state. For this purpose a set of dispersing additives is introduced into the oil. When the diesel engine operates under moderate or high loads the temperature in the combustion chamber is very high, viz. 500–600 °C. This results in oil film oxidation on the pistons and cylinder walls to a greater extent than it occurs in gasoline engines. To improve the oil oxidation resistance, oxidation inhibitors and alkali additives are introduced into the oil. Fuel is the other factor affecting strongly the quality of the diesel oil. Sulfur content is much higher in the diesel fuel than in gasoline one. In combustion sulfur forms a great amount of strong sulfuric acid and to neutralize its effect alkali additives are introduced.

Characteristics of Oil Performance. Presently the oil service life is commonly assigned by the standards set by the car run or operation period. These standards set by the manufacturers are based on the engine operation under normal conditions. Thus, scheduled servicing of AUDI diesel cars including oil replacement are conducted every 7.5 thousand km of run. The manufacturer of Ford diesel cars recommends replacing the oil after 10–15 thousand km of run.

In normal engine operation oil ages gradually and its efficiency is commonly evaluated by the limits of quality characteristics. In practice, viscosity always enters any group of individual characteristics. It depends essentially on oil contamination with insoluble products and fuel, as well as on dispersing-stabilizing properties (DSP) of the oil. Consequently, to adequately interpret variations in oil viscosity one should evaluate not only fuel concentration in it but also the degree of its contamination and DSP.

Oil efficiency is governed by the additives; therefore, in order to evaluate oil quality the activity of additives is monitored. To evaluate the efficiency of dispersing-stabilizing and partly detergent additives DSP of the operating oil is used as a limit parameter. To evaluate anti-wear and neutralizing additive efficiency pH value is used. The latter is also a limit parameter to determine strong acids in the oil whose presence is not permitted since they activate the corrosive wear of engine parts.

Therefore, to evaluate the oil quality the following characteristics are presently used: viscosity, concentration of cooling fluid and fuel in it, DSP, total contamination and pH. Table 4.1 presents the main oil quality characteristics for heavy-duty diesel engines.

Soot Content as a Main Characteristic of Modern Engine Quality. The stringent demands to the monitoring of nitrogen and soot exhausts have resulted in particular the use of exhaust recycling. In this case the exhaust comes to the combustion chamber producing repeated opportunity for the soot to be captured by oil. As a result, the soot concentration in oil increases greatly and causes a number of problems, i.e. DSP deterioration, sludge formation, loss of anti-wear effect, shutting-off the oil flow, viscosity increase, and filter choking [10]. Increased soot content produces more severe conditions of dispersant action.

Lubrizol Co. (USA) has evaluated the effect of different soot concentration on wear of the engine parts. For example, in case of the roller follower the increased soot content affects insignificantly the wear rate as long as particles are small, so this rate is governed mostly by the soot dispersity.

Physical-Chemical Properties of Soot Particles. It has been found in [11] that the soot in the diesel oil contains from 80 to 95 wt% of carbon. Electron microscopy examination of soot particles extracted from the diesel oil shows the initial particle size (before agglomeration) to be 30–50 nm.

Due to small particle size of soot, the used diesel oil can be considered as a colloidal system. Soot particles coagulate under the action of Van der Waals forces of attraction between the particles at certain distance. The main part of the dispersant to be introduced is to produce the adsorption layer of a certain thickness to

Table 4.1 Main characteristics of oil performance in heavy-duty diesel engines

Oil condition	Fuel content in oil, wt%	pH	Water content in oil, wt%	Viscosity variation, %
Effective	No	≥ 6	–	–
Permissible	≤ 3	5.5–4.5	≤ 0.2	≤ 25–30
Pre-failure	≤ 7	–	–	–
Failure	>7	4 and less	1.0	–

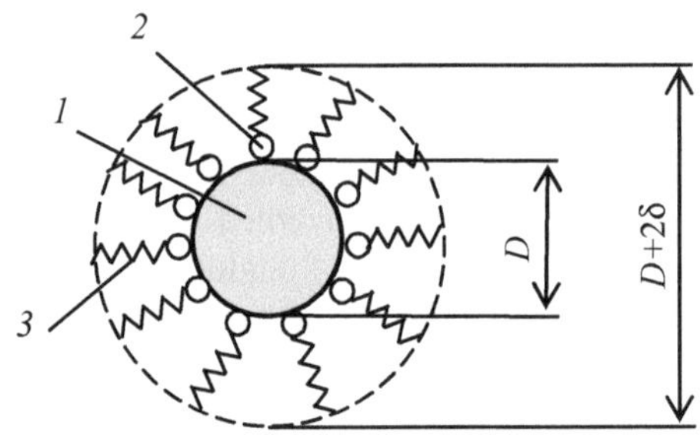

Fig. 4.1 Model of adsorption of molecules on solid particle: *1* solid particle with diameter *D*; *2* polar groups, *3* hydrocarbon chains

reduce forces of mutual attraction between the particles. For this purpose the substances are used whose molecules consist of a short reactive group of atoms and a long hydrocarbon chain.

Let us assume that at the Brownian motion of the particles the least distance between their surfaces is 2δ where δ is adsorption layer thickness (length of adsorbed molecule), i.e. no interaction between the adsorption layers occurs. Figure 4.1 shows schematically the particle *1* of diameter *d* coated with the adsorption layer [12]. Molecules of the surfactant have the polar group *2* bound with the particle surface physically or chemically. A long-chain tail end (hydrocarbon chain) of molecule *3* located in the liquid base is subjected to the Brownian motion.

The thickness of the adsorption layer δ has to be sufficient for energy of attraction between the particles E_v not to exceed the Brownian motion energy kT:

$$|E_v| \leq kT.$$

This means that disordering effect of the Brownian motion prevents the particles from agglomeration. Boundary $|E_v|/kT = 1$ divides the regions of aggregate-stable fluid and unstable one whose disperse phase coagulates when solid particles collide.

Laboratory studies have shown that in case of well dispersed soot in Mack six-cylinder diesel engine the average soot particle size is about 150 nm and the maximal size is a little larger than 350 nm [11]. But in case of poorly dispersed soot the average particle size is about 210 nm while the maximal size exceeds 500 nm.

4.3 Laboratory Methods of Monitoring Soot Content in Oil and Its Properties

Presently thermal gravimetric and infrared analysis, sedimentation technique, paper chromatography, and photon correlation spectroscopy (Fig. 4.2) are used in research laboratories to measure the soot particle content [13].

Thermal gravimetric analysis (*TGA*) [11] is a widely used method providing the most accurate evaluation of soot content as a mass percentage. The measurement

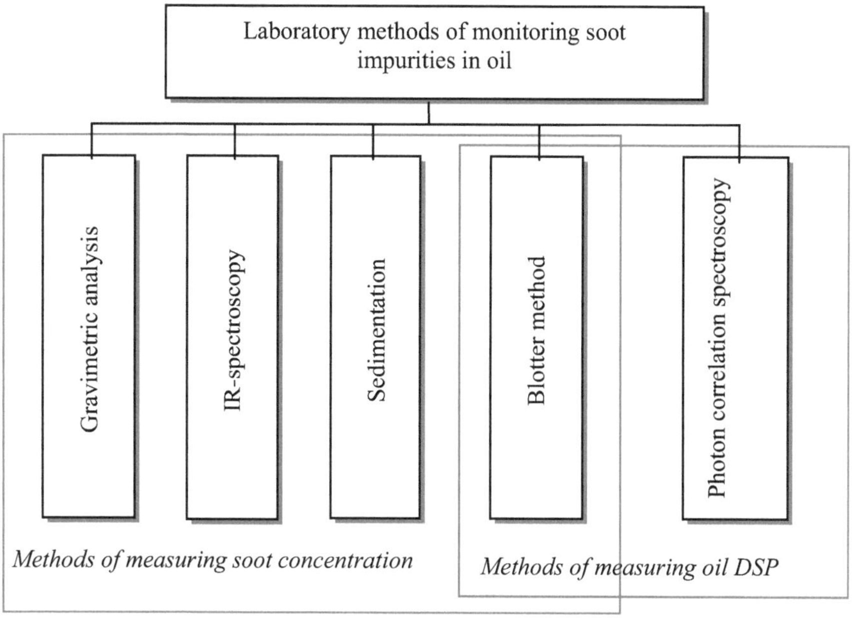

Fig. 4.2 Classification of laboratory methods of monitoring soot impurities in oil

comprises two stages. The first one involves the gradual heating of the sample in the nitrogen medium to evaporate volatile fractions. The evaporation usually occurs at 600–650 °C. At the second stage nitrogen is replaced by oxygen and the temperature elevates up to 750 °C to oxidize the oil. The soot concentration is evaluated as a sample mass loss at the second measurement stage. Though the method is accurate enough it can be only used under laboratory conditions. In addition, it is very expensive and one analysis costs around $200.

Infrared analysis (IR) has become very popular to measure the soot concentration [14]. Its results correlate well with those of TGA. It's relatively cheap due to rapid growth of a number of laboratories dealing with Fourier Transform Infrared Spectroscopy (FTIR). Maximal soot absorption occurs in the region close to wave number $k = 2000$. The range is relatively free of superposition of some other components of the oil or impurities which makes the soot measurement a relatively simple technique. To record spectra of diesel oil as a strongly absorbing fluid the method of disturbed total internal reflection is optimal. It is based on the absorption of radiation in the oil surface layer being in contact with the optical surface. The main disadvantage of the method is a strong increase of infrared radiation absorption with increasing the soot particle size.

Sedimentation method comprises the separation of insoluble substances from the oil after diluting it with pentane or toluene [10]. The insoluble substances are removed by the high-speed centrifuging or filtering through a membrane filter. After centrifuging the content of the insoluble substances is evaluated either by mass or by volume. The mass of a new filter is subtracted from the mass of the used filter to determine the mass of the insoluble substances. The method is relatively simple and cheap. Its main disadvantage is in the fact that all insoluble substances are weighed altogether. The solvent also affects the measurement result. If pentane is used then organic oxides are involved in the measurement of all insoluble substances. Toluene keeps organic oxides insoluble.

The Blotter method [10] is a method that involves just one or two oil drops on the standard chromatographic paper. Insoluble substances are separated from the oil and provide fast visual identification of soot by the chromatogram. Further the paper can be dissolved by different solvents which provide the indication of insoluble substances. The method of paper chromatography can also be used to evaluate oil DSP by the characteristic rings on the chromatogram. The method is enough simple though it does not yield quantitative evaluation.

Photon correlation spectroscopy, a specific method using dynamic (quasi-elastic) scattering of laser radiation allows one to evaluate the size distribution of soot particles contained in oil. The method is based on temporary variation of scattered optical radiation by the fluid containing suspended particles [15]. The intensity of scattered radiation depends on such characteristics as refractive index, particle size, and concentration, particle displacement being the critical one. The particle displacement caused by Brownian motion is inversely proportional to the particle size and it is a random function for micron and submicron ranges. Applying autocorrelation analysis of radiation intensity one can evaluate the size distribution of particles. In this case it is assumed that particles are of spherical shape so as to eliminate the dependence of the results on the density and refractive index of the particles. The method of photon correlation spectroscopy is rather complex. In addition, because of uncertain shape of typical aggregates the measurement of particle size using this method implying spherical particle shape is not adequate. However, the non-spherical shape of particles could be partially compensated if measuring scattered radiation at an angle of $< 90°$.

All the methods outlined above are laboratory ones and can not solve completely an increasing problem of the soot content monitoring to determine the optimal time of oil replacement.

4.4 Portable and Built-in Devices for Monitoring Soot in Oil

Portable monitoring devices. Nowadays an IR Soot Meter (Spectro Scientific Co.) has been developed based on IR-spectroscopy that uses the disturbance of the law of total internal reflection. It is intended to monitor the soot concentration in oil under field conditions [16]. The instrument comprises a cell whose bottom is made of measuring optical plate of cubic zirconia. The cell is filled with the oil to be tested with a pipette. The measurement results are presented as soot mass concentration in the oil. The device weight is <2.5 kg and overall dimensions are

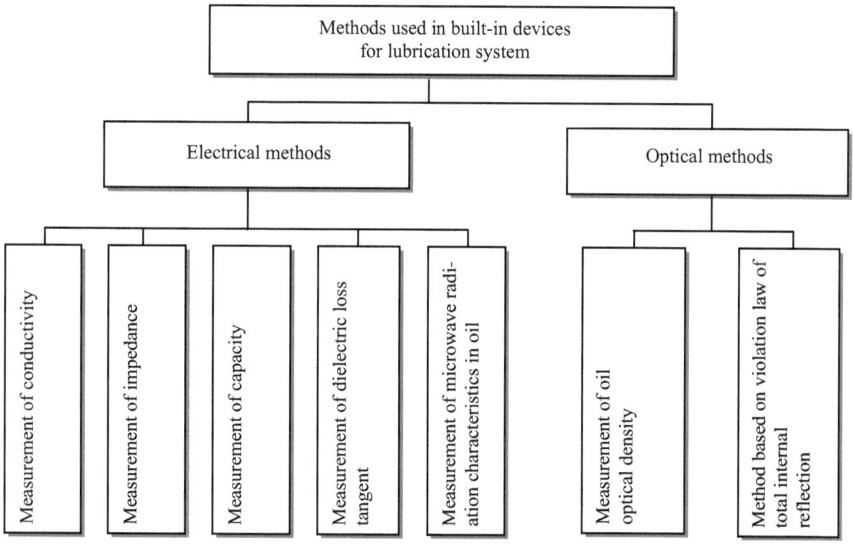

Fig. 4.3 Classification of methods for monitoring soot content used in devices built in the lubrication system

<150 mm^3. The laboratory tests performed in the Mobil's Paulsboro Technical Center with engines Mack T-8 and DDC Series 60 have shown a good correlation between the Soot Meter results and gravimetric analysis results.

Monitoring devices built in the oil circulation system. To realize on-line soot content measurement for vehicles it is required to produce a measuring system which is relatively cheap to be installed to large number of cars and fairly reliable to operate in the working medium of a diesel engine.

Soot contained in oil varies its electrical (conductivity and permittivity) and optical properties (optical density). Therefore, available devices built in the car lubrication system are based on electrical and optical methods (Fig. 4.3).

Electrical methods. Pure mineral oils have a very high electric resistance and relatively low permittivity. These electric properties vary with oil deterioration and contamination. In particular, increasing soot content reduces considerably specific resistance ρ and increases permittivity ε of the oil. The model system of measuring electrodes *1* and *2* with contaminated oil *3* between them can be presented as a capacitor whose capacity C depends on the dielectric permittivity. It has a parallel resistor whose conductivity g depends mainly on the soot particle concentration (Fig. 4.4).

Total electric conductivity is determined as

$$\gamma = \sqrt{g^2 + (2\pi f C)^2}, \tag{4.1}$$

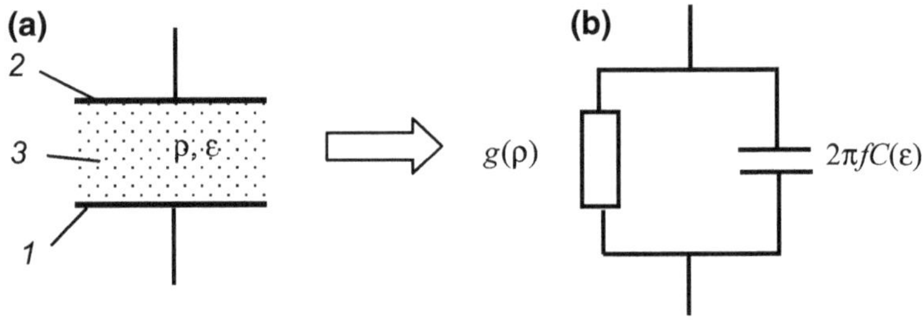

Fig. 4.4 System of electrodes *1* and *2* with contaminated oil *3* between them (**a**) and its electrical model (**b**)

where $C = \varepsilon S/d$, ε—oil permittivity, *S*—capacitor surface area, *d*—gap between the plates, *g*—conductivity of the cell, and $2\pi fC$ is capacitance component of the total conductivity.

Methods based on oil electric properties as a function of its contamination can be divided on those measuring Ohm's resistance (conductivity), total resistance (impedance), capacity, dielectric loss tangent, and characteristics of ultrahigh frequency (UHF) electromagnetic radiation. The electrical methods are realized in the devices built in the lubricating system whose schematic is shown in Fig. 4.5.

Initially an attempt was made to use the apparatus for detection of oil the contamination based on the measurement of Ohm's component of the oil resistance [17].

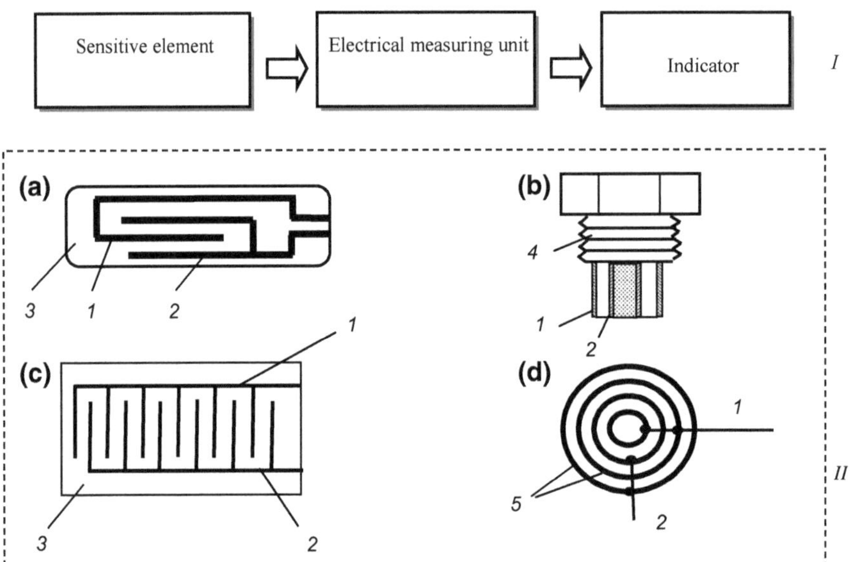

Fig. 4.5 Schematic diagram (*I*) of electric devices for monitoring soot content in oil: *a, b, c, d*—variants of realization of sensitive element (*II*); *1* and *2* measuring electrodes; *3* substrate; *4* insert; *5* concentric cylinders

The apparatus comprises a sensitive element (Fig. 4.5a) including sensitive electrodes *1* and *2* deposited onto the substrate *3*. The electrodes are connected to the electric resistance gauge unit (Ohmmeter). An indicator (signal lamp) is connected to the electric unit. With increasing soot concentration in the oil its conductivity (g) increases, i.e. the resistance between the measuring electrodes decreases. The characteristics of the electric unit circuit are chosen to provide the current in the measuring circuit to be sufficient for the lamp to be lighted at a critical oil contamination.

However, measurement of the resistance using the direct current is affected considerably by the electric noise. Therefore, methods and devices have been suggested whose operation is based on the measurement of the total resistance (impedance) using the alternating current [18–20]. This allows one to perform frequency filtration of the data signal. In this case the conductivity of the measuring sell is governed by the total conductivity (Formula 4.1).

Thus, the operation of the oil quality sensor [18] intended for measuring impurity content in oil is based on the measurement of the oil total conductivity that is governed by the degree of oil contamination using the alternating current of frequency 100–1000 Hz. Concentric cylindrical sensor electrodes *1* and *2* (Fig. 4.5b) are separated by a gap sufficiently wide for free oil circulation and measurement of the conductivity between the electrodes. The measuring electrodes installed into the engine crank case through the insert *3* are connected to the unit of conductivity measurement and comparison with limit level of the signal. With exceeding the limit, the warning signal is actuated on the indicator panel of the car.

In the other measuring device for monitoring the total conductivity [19] two overlapping combs *1* and *2* deposited onto the substrate *3* are used (Fig. 4.5c). The gap between the electrode elements falls within a range from 0.12 to 1 mm. In the given instrument the measuring signal frequency exceeds 1 Hz.

To decrease the temperature dependence of the output signal caused by the Ohm's component of the conductivity the methods are suggested based on using the high frequencies. In this case the capacitive conductivity exceeds considerably the Ohm's one. Therefore, the total conductivity is governed by its capacitance component. The latter is proportional to the permittivity of the oil depending on the soot concentration (Formula 4.1).

In one of these instruments suggested by Yasuhara [21] the alternating voltage of frequency 50–500 kHz is applied to the sensitive capacitor. The instrument comprises a couple of capacitor electrodes coated with the insulating material and immersed in the oil. Either two parallel plates or two sets *1* and *2* of concentric cylinders *5* are used as the electrodes (Fig. 4.5d). Voltage at the capacitor output after processing in the measuring unit is applied to the indicator panel.

Some other devices on the basis of the measurement of capacitive resistance are known. The devices are reported which can simultaneously detect and identify the corrosion products, impurities, ferromagnetic, and non-ferromagnetic particles in oil [22], as well as the built-in analyzer for continuous monitoring of the diesel engine lubricating system [23]. Sensitive element *1* (Fig. 4.6) of these devices is made as the capacitor formed by two concentric systems *2* and *3*. Each of them

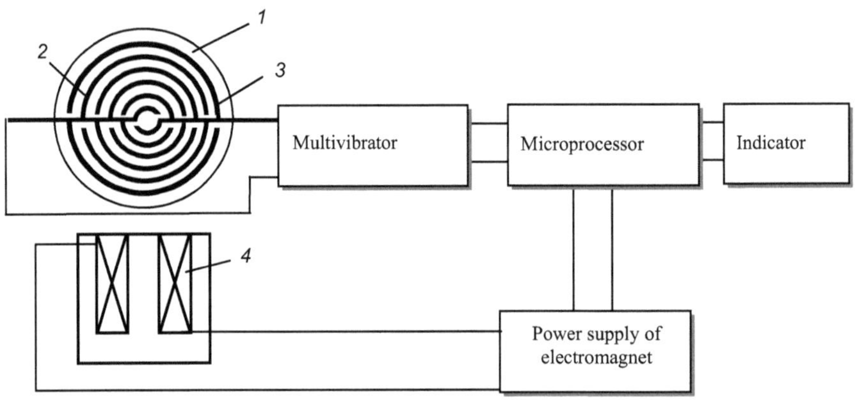

Fig. 4.6 Built-in analyzer of lubricating system of diesel engine: *1* sensitive element; *2* and *3* concentric systems of conductors; *4* electromagnet

presents a set of wires having diameter 250 μm and forming semi-circles. The sensitive element is installed in parallel with the resistor thus forming an RC-circuit that is included in the multi-vibrator circuit and governs the frequency of the output signal to be generated. After signal frequency transformation into voltage in the microprocessor circuit the data signal is applied to the indicator. In addition, the device contains electromagnet *4* to trap, orient and hold ferromagnetic particles on the surface of the sensitive element. Thus it is possible to obtain additional information about ferromagnetic wear particles content in oil.

Different variants of connecting the capacitive sensitive element and signal processing circuits are used to make the oil monitoring devices more reliable.

Thus, in the monitoring device WO9850790 [24] the sensitive capacity element whose configuration is similar to that of the sensitive element shown in Fig. 4.5b and the inductive element are connected in parallel in the vibrator circuit that operates at a frequency of 10–300 MHz. The amplitude and frequency of vibrator oscillations depend on the capacity of the sensitive element and hence on the oil permittivity governed, in particular, by the soot concentration. The amplitude of the output signal is advisable to use as an informative characteristic. It depends on the dielectric loss tangent of the oil which increases significantly the device sensitivity.

The dependence of the oil permittivity on the soot particle concentration has become the basis of the method for monitoring soot content of the oil in the diesel crank case. It involves measurement of the wavelength of microwave electromagnetic radiation within the 8–12 GHz range independent of the oil composition (additives and other impurities) [25]. The instrument comprises Gunn diode *1* as the oscillator (Fig. 4.7) operating at frequency 10 GHz. The oscillator is connected coaxially to the end of the transmission line having inner *2* and outer *3* conductors. The second end of the line is connected to the probe *5* having the same inner *2′* and outer 3′ conductors made from a fine mesh. The bottom of the probe 5 is shorted by screen 6, which is a conductor such as a mesh. Window *4* made of a dielectric

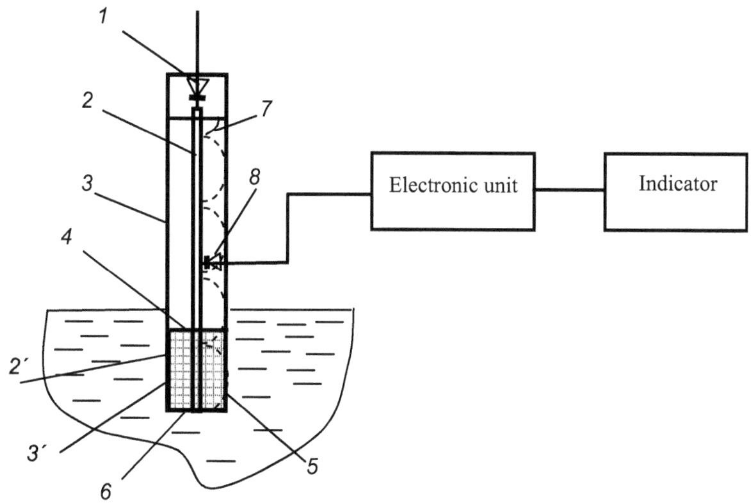

Fig. 4.7 Device for detecting soot content in the oil contained in diesel crank case by measuring microwave radiation wavelength: *1* Gunn diode; *2, 2′* and *3, 3′* inner and outer conductors; *4* window; *5* probe; *6* screen; *7* standing wave; *8* detector

material keeps the oil from entering the transmission line. Microwave radiation is transferred to the probe immersed into the oil. Incident radiation and radiation reflected from the screen *6* generates the standing wave *7* in the transmission line. The radiation wavelength depends on its frequency and relative permittivity of the medium. When varying the permittivity with increasing soot content the wavelength changes and wave node position is shifted along the transmission line axis. Detector *8* located at the node at the zero soot concentration is used to measure the variation in the radiation wavelength. This wavelength varies in inverse proportion to the square root of relative permittivity of the oil. Therefore, the soot content in the oil is evaluated by the radiation wavelength variation.

The major shortcoming of the methods based on the dependence of the oil permittivity on the contamination in measuring the soot concentration is that the permittivity depends significantly on other impurities, mainly water, cooling fluid and fuel. The relative permittivity of the oil commonly varies from 1.6 to 3.2 depending on its grade and operation time. For example, the permittivity of the engine oil can increase from 2.19 to 3.2 after 400 h of its usage. The relative permittivity of water and cooling fluid are about 87.5 and 37.5, respectively. These impurities when entering the oil increase considerably its permittivity. Moreover, the permittivity depends greatly on the temperature. Hence, the devices of temperature compensation are needed.

The content of water, cooling fluid and fuel affects less when measuring the soot concentration with optical methods since soot has high absorption coefficient, so the absorption of optical radiation is mainly governed by the soot content. Soot weight concentration c in oil can be evaluated by the following relationship:

$$c = -\frac{1}{K_{ext,\lambda} l} \lg \tau, \tag{4.2}$$

where $K_{ext,\lambda}$—extinction coefficient at wavelength λ, l—optical path length in oil, τ—transmittance (ratio of the intensity of radiation passed through the medium to that of incident radiation).

When developing the devices monitoring the soot concentration the main problem is to provide a short optical path ($l < 400$ μm) in the oil at which radiation is not absorbed entirely, so it is necessary to provide the constancy for the optical path and oil passage through a narrow gap. One of the solutions is suggested in the photometric device [26] (Fig. 4.8a), where optical windows *1* and *2* are made in the form of plate-convex lenses whose surfaces contact each other in the oil. Optical radiation produced by the light-emitting diode *3* is focused to photodetector *4*. After amplification the electrical signal of the photodetector is transmitted to the voltage indicator; the voltage is recalculated into the soot concentration in the oil using the calibration plot (Fig. 4.8b).

Yet even more reliable method to provide a constant optical path in the oil is realized in the device based on the absorption of radiation in the layer of thickness t (Fig. 4.9a) during its total internal reflection from the interface *4* (Fig. 4.9b) between the optical material with refractive index n_1 and the oil under testing with refractive index n_2 ($n_1 > n_2$) [27]. The measurement technique is similar to the technique used in the aforementioned IR Soot Meter.

The device comprises the emitting diode with collimating system *1*, measuring prism *2*, measuring photodetector *3*, and photodetector *5* used to compensate the fluctuation.

In total internal reflection the optical radiation penetrates into the less dense medium at a distance t, which can be determined by the formula:

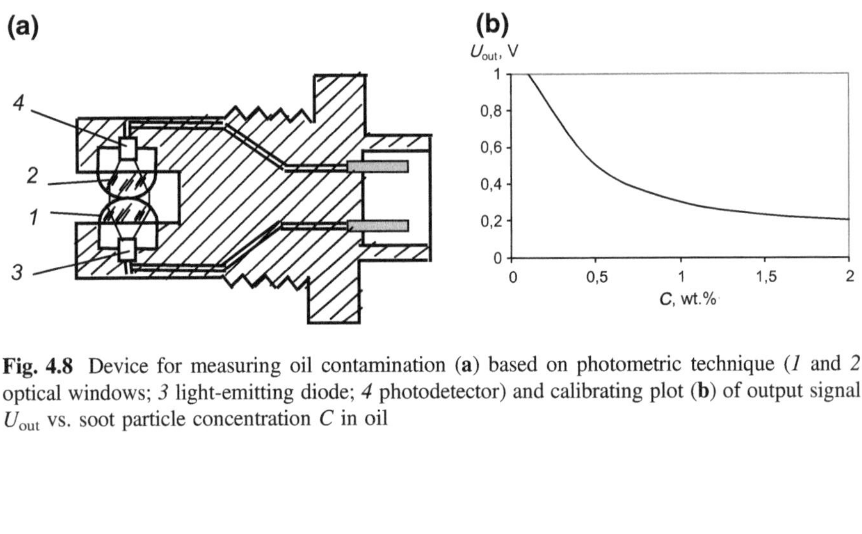

Fig. 4.8 Device for measuring oil contamination (**a**) based on photometric technique (*1* and *2* optical windows; *3* light-emitting diode; *4* photodetector) and calibrating plot (**b**) of output signal U_{out} vs. soot particle concentration C in oil

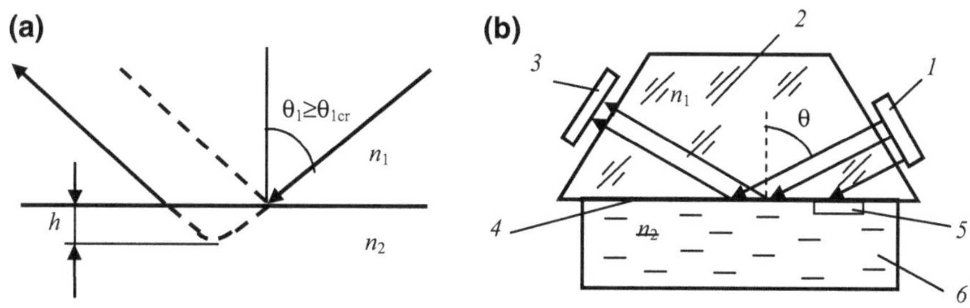

Fig. 4.9 Scheme of penetration of optical radiation into the test oil (**a**) when violating the total internal reflection law (θ_1—angle of incident of optical beam to the interface, θ_{1cr}—critical angle of incident) and device for monitoring oil deterioration (**b**) based on violation of total internal reflection law: *1* light-emitting diode with the collimating optical system; *2* measuring prism; *3* and *5* measuring and compensating photodetectors, respectively; *4* the surface of photodetector; *6* test oil

$$h = \frac{\lambda}{n_1^2} 2\pi \left(n_1^2 \sin^2 \theta_1 - n_2^2\right)^{1/2}. \tag{4.3}$$

Taking $l = 2h$ one can evaluate the soot concentration in oil by Formula (4.2). As is seen, the device doesn't need the gap between the windows to perform photometric measurements, i.e. the problem of oil circulation in a narrow gap is also solved.

The shortcoming of both optical and electrical methods results from the contamination of the sensitive element surfaces, thus with time records are invalid. To decrease the surface contamination it is proposed to use the silicon coating on electrodes [28] or Teflon coating (AF-2400) on optical surfaces [29]. However, it was no success in coating use and the problem of surface protection needs further attack.

4.5 Fiber–Optic Sensors of Soot Content in Diesel Oil

Most challenging way of development of the new methods and means for monitoring of soot content in diesel oil is the usage of fiber optic detectors, which are not impacted by electrical noise and disturbance. Besides, informative parameter of optical detectors is light adsorption depending mainly on soot content in oil in contrast to electrical (capacitive) devices, where measured electrical parameters depend on complex of factors (content of soot, water, acids, metal impurity etc.).

Basics of development of fiber-optic detectors with sensitive light guide for oil condition monitoring are described below. Operation of such detectors is based on the multiple total internal reflection of the light from interface of light guide—external medium (oil) and the change of radiation intensity when parameters of test oil vary.

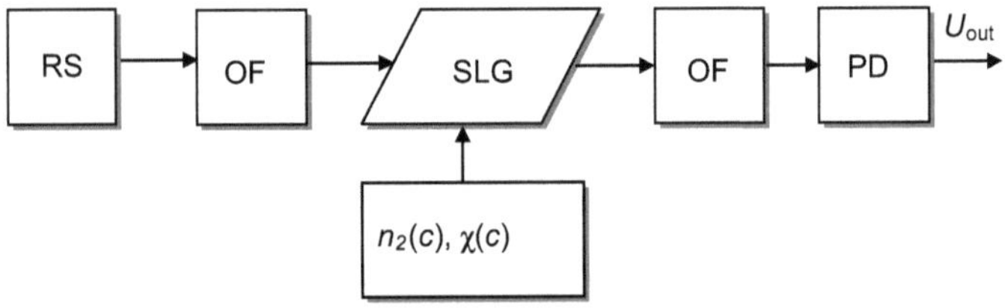

Fig. 4.10 The general scheme of detector with light guide: radiation source (RS), sensitive light guide (LG), optical fibers (OF), and photodetector (PD)

Light from the radiation source is entered through the optical fiber to sensitive light guide, which surface is in the contact with the test oil. Intensity of radiation passed the sensitive light guide is varied when optical properties of the oil change (Fig. 4.10).

Optical properties of the test oil are described by complex refractive index:

$$\bar{n}_2 = n_2 - i\chi.$$

The real part n is the refractive index, which equals to ratio of velocity of the wave traveling through the oil to velocity in vacuum The imaginary part χ is absorption index, which defines reduction of radiation intensity as a result of medium absorption.

When soot concentration in oil increases the absorption index χ and refraction index n_2 are changed. As a result the intensity of radiation passed through the sensitive light guide decreases. Modulated optical radiation is passed through the optical fiber to the photodetector converting the optical signal to electrical voltage U_{out}.

The sensitive light guide is a flat optical substrate. The optical radiation P_{in} generated by the radiation source (RS) with aperture angle φ_{ap} enters into the light guide 3 with refractive index n_1 by prism 1 (Fig. 4.11). The radiation propagation in the light guide occurs by reflections from the interfaces of the light guide—

Fig. 4.11 Optical diagram of a flat light guide: P_{in}, P_{ref}, P_{refr}, P_{inc} and P_{out} are input, reflected, refracted, incident and output powers correspondingly

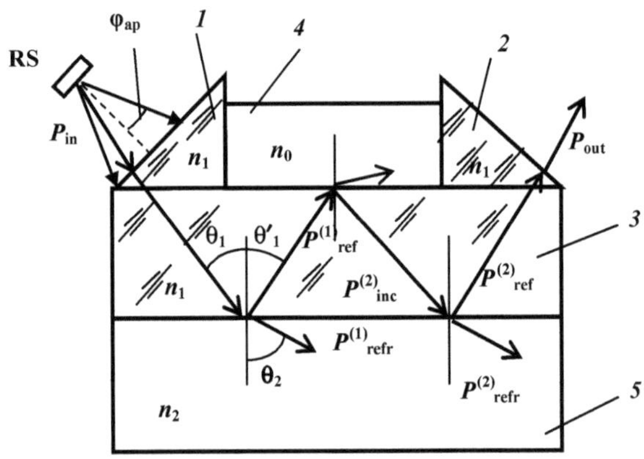

substrate 4 with refractive index n_0 and the light guide—external medium (test oil) 5 with refractive index n_2.

In the light guide the aperture angle φ_{ap} of radiation source is transformed into the aperture angle with limit angles $\pm\,\theta_{1max}$.

There are three key factors causing loss of radiation power at light guide outlet during optical properties variation of test oil: total internal reflection, attenuated total reflectance, and reflection of rays violating TIR condition.

Total Internal Reflection

Rays incident on light guide—external medium interface at an angle exceeding the critical angle of TIR, i.e.

$$\theta_1 \geq \theta_{1cr} = \arcsin(n_2/n_1),$$

undergo TIR and pass through the light guide practically without power losses.

Increase of refractive index of external medium causes θ_{1cr} increase that decreases the part of RS rays for which the TIR law is satisfied. Owing to this fact the power of radiation P_{out} at LG outlet is decreased. The value of reflection coefficient at TIR exceeds its most high values at mirror reflection from polished surfaces and is practically equal to 1 with high accuracy. Besides, this index does not depend on light wavelength and even at multiple TIR the spectrum of complex radiation does not vary.

The relationship between the critical angle and the refractive index of external medium can be used in detector of oil deterioration in case when oxidation or contamination of oil cause variation in oil refractive index. For example if refractive index of test oil changes from $n_2 = 1.47$ (clean oil) till *1.51* (used oil), then critical angle changes in *3°* (Fig. 4.12) that can easily be measured.

Attenuated Total Reflection.

At ideal TIR conditions the reflection coefficient from interface of the light guide—test oil due to the losses in this case are caused by absorption in the medium of light

Fig. 4.12 Critical angle θ_{1cr} dependence on refractive index of external medium n_2 (test oil) at refractive index $n_1 = 1.81$ of the sensitive light guide

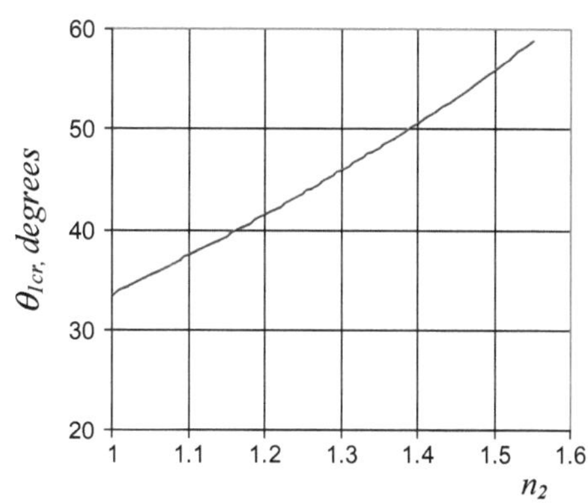

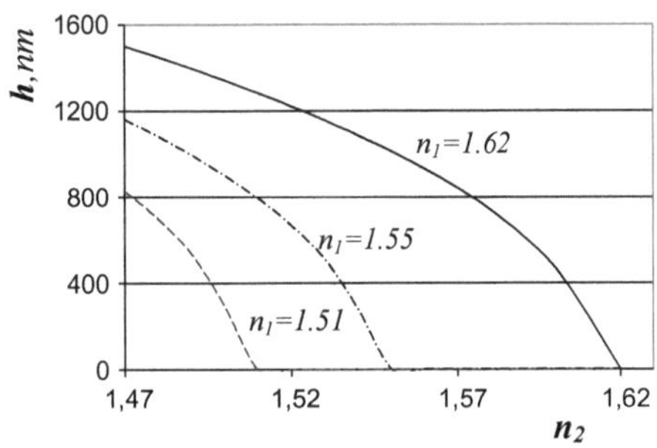

Fig. 4.13 The penetration depth h as function of oil refractive index n_2 for light guide refractive indexes *1.51, 1.55* and *1.62* (ray incident angle $\theta_1 = 85°$)

guide element. However in real conditions the light penetrates into the thin layer of the test oil absorbing the optical radiation. That is, the violation of the total internal reflection takes place and reflection coefficient decreases. At the attenuated total reflectance (ATR) electromagnetic field penetrates into the medium with less optical density ($n_2 < n_1$) to specific depth h (Fig. 4.9a) depending on the light wavelength λ and its amplitude is exponentially attenuated with distance from the interface. The depth of penetration can be estimated by the Formula (4.3).

Figure 4.13 shows the penetration depth as a function of oil refractive index at incident angle $\theta_1 = 85°$ for three light guides having different refractive indexes *1.51, 1.55* and *1.62*. The higher is the difference between refractive indexes of oil and light guide material, the higher is the depth of penetration.

Absorption of the light radiation passed through the light guide depends on both the soot content in oil and the penetration depth (refractive index of the light guide). Figure 4.14 presents the dependence of relative output P_{out}/P_{in} on soot content at incident angle $\theta_1 = 85°$ for light guide refractive indexes 1.51, 1.55 and 1.62. The dependences are calculated basing on Lambert-Beer's law:

$$P_{\text{out}} = P_{\text{in}} \exp(-K_{ext,\lambda} l), \tag{4.4}$$

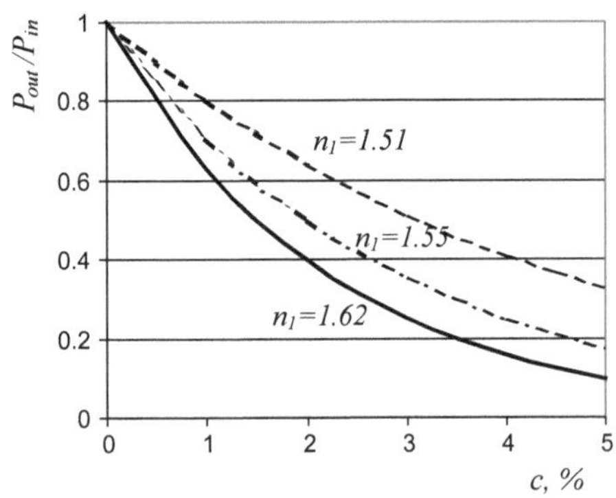

Fig. 4.14 Value P_{out}/P_{in} as a function of carbon soot concentration at incident angle $\theta_1 = 85°$ for light guides with refractive indexes *1.51, 1.55* and *1.62*

where l is the optical length in oil, $l = 2hm$ (m—number of reflections); $K_{ext,\lambda}$ is an extinction coefficient at wavelength λ. The extinction coefficient equals to the sum of natural coefficients of absorption and scattering. In case of soot particles analysis the scattering is considerably less than absorption, therefore extinction coefficient equals to the natural coefficient of absorption $K_{ext,\lambda} = \alpha_{\lambda}$. This coefficient can be presented as $\alpha_{\lambda} = k_{\lambda}c$, where k_{λ}—coefficient independent on soot concentration c and characterizing the interaction between molecule of absorbing medium and optical radiation at wavelength λ. In case of oil $k_{\lambda} \approx 40\frac{1}{\text{MM}}$ at the red wavelength of the radiation.

The reflection coefficient of light from media interfaces becomes less than unity owing to absorption of light in reflecting medium layer where the wave penetrates.

Equation (4.4) takes into account absorption only but leaves out of account the reflectivity changes.

Reflectivity of interface between two mediums is defined by Fresnel's formulas. In the case when one of mediums is absorbing the reflection coefficient can be estimated by the following approximated expression [30]

$$R(\theta_1) = 1 - \frac{4 \cdot n_1 \cdot n_2 \cdot \cos\theta_1}{(n_1^2 - n_2^2)(n_1^2 \cdot \sin^2\theta_1 - n_2^2)^{1/2}}\chi, \tag{4.5}$$

where χ is absorption index of external medium (imaginary part of complex refractive index $\bar{n}_2 = n_2 - i\chi$).

Equation (4.5) is valid for incident angles greater than critical $\theta_1 \geq \theta_{1cr}$ and for $\chi \leq 0.004$.

At increasing number of reflections till m the reflectivity is defined as:

$$R_m(\theta_1) = R^m(\theta_1) = P_{out}/P_{in}. \tag{4.6}$$

Calculations have been performed for optical glass ($n_1 = 1{,}51$) and oil ($n_2 = 1.48$) interface and χ values of 0.001, 0.002 and 0.004. The results are shown in Fig. 4.15 for one ($m = 1$) and four ($m = 4$) reflections.

It is clear that reflectivity significantly depends on medium absorption at several reflections. This dependence can be used as basis for development of carbon soot detector.

In order to analyze the connection between the optical glasses with different refractive indexes and variations in oil refractive index the reflectivity $R_m(\theta)$ have been calculated for three refractive indexes of glass ($n_1 = 1.51$, *1.55 and 1,62*) assuming that $\chi = 0.001$. It's clear from Fig. 4.16 that if the light guide refractive index is close to refractive index of the test medium, then the light guide is more sensitive ($n_1 = 1.51$; $m = 4$).

Reflection of rays violating TIR condition. Rays incident on the light guide—external medium interfaces at an angle less then critical angle of TIR i.e. $\theta_1 \leq \theta_{1cr} = \arcsin(n_2/n_1)$ undergoes multiple reflections.

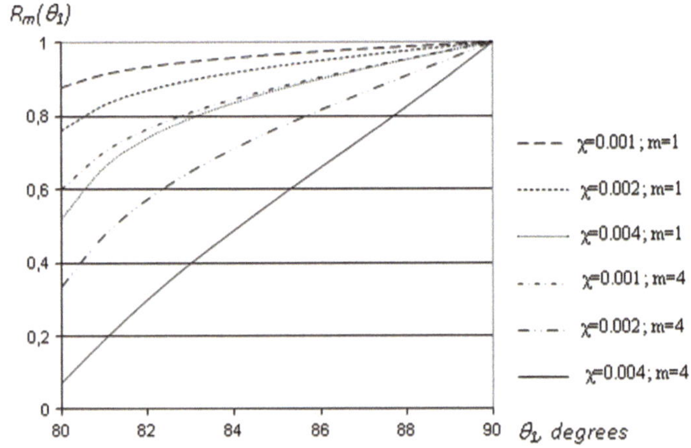

Fig. 4.15 Reflectivity $R_m(\theta_1)$ of optical glass/oil interface as function of incident angle θ_1 for different values of absorption index χ and number of reflections m

Fig. 4.16 Reflectivity $R(\theta_1)$ of optical glass/oil interface as function of incident angle θ_1 for different refractive indexes of optical glass at value of absorption index χ = *0.001*

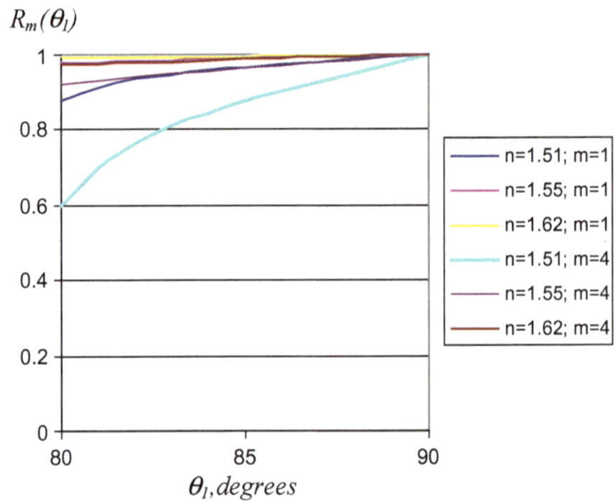

The power of the ray reflected from the light guide—external medium interface $P^{(1)}_{ref\perp}$ in case when polarization plane of ray is perpendicular to incident plane is defined by the expression

$$P^{(1)}_{ref\perp}(\theta_1, n_2) = R_\perp(\theta_1, n_2)P^{(1)}_{inc\perp}(\theta_1),$$

where $P^{(1)}_{inc\perp}(\theta_1,\ n_2)$—power of incident radiation, $R_\perp(\theta_1,\ n_2)$—reflectivity of light guide—external medium interface when polarization plane of the ray is perpendicular to the incident plane.

Similarly when the ray polarization plane is parallel to the incident plane the power of the reflected rays can be estimated as

$$P^{(1)}_{ref//}(\theta_1, n_2) = R_{//}(\theta_1, n_2) P^{(1)}_{inc//}(\theta_1),$$

where $R_{//}(\theta_1,\ n_2)$—reflectivity of light guide—external medium interface when polarization plane of the ray is parallel to the incident plane.

For natural light we have the following relation

$$P^{(1)}_{ref}(\theta_1, n_2) = 0,5[R_{\perp}(\theta_1 . n_2) + R_{//}(\theta_1, n_2)] P^{(1)}_{inc}(\theta_1).$$

Reflectivity according to Fresnel's formulas can be written as follows:

$$R_{\perp}(\theta_1, n_2) = \left[\frac{n_1 \cos\theta_1 - \sqrt{n_2^2 - n_1^2 \sin^2\theta_1}}{n_1 \cos\theta_1 + \sqrt{n_2^2 - n_1^2 \sin^2\theta_1}} \right]^2,$$

$$R_{//}(\theta_1, n_2) = \left[\frac{n_2^2 \cos\theta_1 - n_1 \sqrt{n_2^2 - n_1^2 \sin^2\theta_1}}{n_2^2 \cos\theta_1 - n_1 \sqrt{n_2^2 - n_1^2 \sin^2\theta_1}} \right]^2.$$

Output power after m reflections from the interface can be written as:

$$P_{out}(\theta_1, n_2) = 0,5[R^m_{\perp}(\theta_1, n_2) + R^m_{//}(\theta_1, n_2)] P^{(1)}_{inc}(\theta_1). \tag{4.7}$$

That is, when refractive index n_2 increases, output power decreases. Output power is equal to zero for any polarization if refractive index n_2 is equal to index n_1.

If the reflecting surface is bounded by external medium of complex refractive index (n_2-$i\chi$), then reflectivity is modified in accordance with the following equations for polarization perpendicular to the plane of incidence

$$R_{\perp}(\theta_1, \bar{n}_2) = \left\{ \frac{[(n_2 - i\chi)^2 - n_1^2 \cdot \sin^2\theta_1]^{1/2} - n_1 \cos\theta_1}{[(n_2 - i\chi)^2 - n_1^2 \cdot \sin^2\theta_1]^{1/2} + n_1 \cos\theta_1} \right\}^2 \tag{4.8}$$

and for polarization parallel to the plane of incidence

$$R_{//}(\theta_1, \bar{n}_2) = \left\{ \frac{n_1[(n_2 - i\chi)^2 - n_1^2 \cdot \sin^2\theta_1]^{1/2} - (n_2 - i\chi)^2 \cos\theta_1}{n_1[(n_2 - i\chi)^2 - n_1^2 \cdot \sin^2\theta_1]^{1/2} + (n_2 - i\chi)^2 \cos\theta_1} \right\}^2. \tag{4.9}$$

In case of a few times of ray reflections from the interface the output power becomes practically equal to zero (Fig. 4.17).

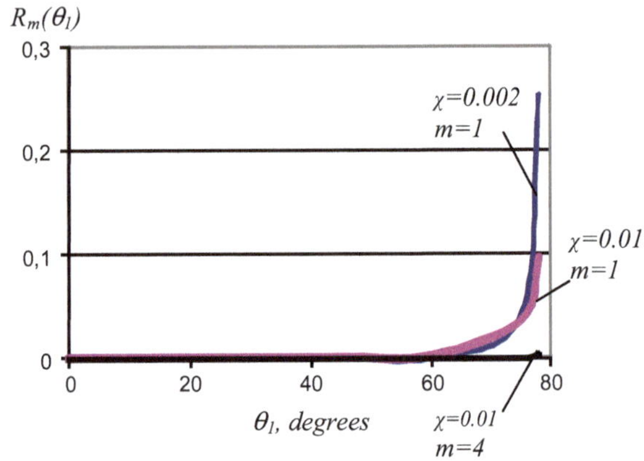

Fig. 4.17 Reflectivity $R_m(\theta_1)$ of optical glass–oil interface as function of incident angle $\theta_1 < \theta_{1cr}$ for different oil absorption indexes χ at values of $n_1 = 1.51$ and $n_2 = 1.48$ and number of reflections $m = 1$ and $m = 4$

Thus, the decreasing of radiation power at the outlet of the optical light guide element during oil degradation is caused by three factors.

(1) Oil oxidation and increase of concentration of fine contaminants (particularly, soot) lead to increase of refractive index n_2. As a result the part of radiation source aperture, which undergo TIR, decreases and output power decreases also.
(2) Besides, oil oxidation and its contamination gives rise to absorption in the thin layer at TIR; that is, reflection index decreases for angles $\theta_1 \geq \theta_{1cr}$ and as a result power decreases at the outlet of light guide.
(3) When oil refractive index increases the reflective index decreases for rays passing through the light guide at an angle less than the critical one and hence, the output power decreases. However, at several reflections this phenomenon is negligible.

It should be noted that variation in refractive index of external medium (n_2) causes variation in aperture of light beam at the outlet of light guide (Fig. 4.18).

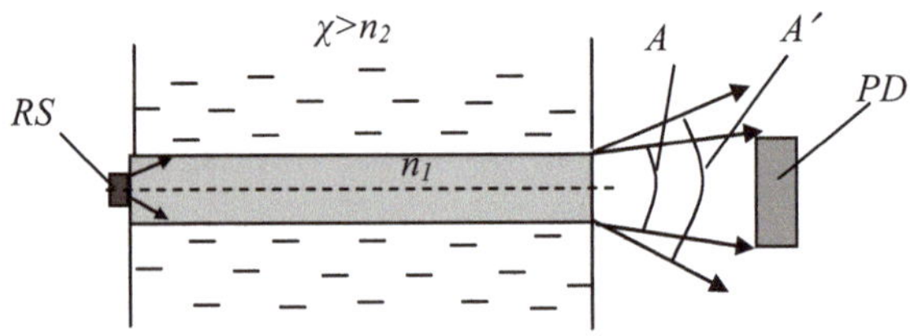

Fig. 4.18 Influence of medium refractive index (n_2 *and* χ) on aperture angle (*A and A*′) of the optical beam at the light guide (n_1) outlet. RS is the radiation source and PD is the photodetector

Therefore, if the sensitive area of photodetector is small (not enough to receive the full optical beam at maximal refractive index of the external medium) then output signal of photodetector will vary. The aperture angle A can be calculated using the formula:

$$A = \arcsin\left(\frac{1}{n_2}\sqrt{n_1^2 - n_2^2}\right).$$

We can see that the less difference between n_1 and n_2, the less aperture angle A, the higher flux intensity on PD area and the higher the output signal. This phenomenon can be observed when the oil ($n_2 \approx 1.48$) replaces the air ($n_2 = 1$).

Estimation of the light guide detector effectiveness

Effectiveness of the light guide detector can be estimated by modulation index M, which is defined by equation [30]:

$$M = (P_{\max} - P_{\min})/P_{\max}, \tag{4.10}$$

where P_{max} and P_{min}—maximal and minimal values of output power at changing properties of external medium. That is the modulation index shows how output signal is changed in relative units.

Powers P_{max} is defined at minimal value of refractive index of the test oil ($\bar{n}_{2\min} = n_{2\min} - i\chi_{\min}$), and P_{min} is defined at maximal value of the refractive index $\bar{n}_{2\max} = n_{2\max} - i\chi_{\max}$:

$$P_{\max} = \int\limits_{\theta_{1\min}}^{\theta_{1\max}} P_{out}(\theta_1, \bar{n}_2) d\theta_1 |_{\bar{n}_2 = \bar{n}_{2\min}}. \tag{4.11}$$

Minimal and maximal values of the refractive index equal to the index of fresh oil and used oil with maximal (limit) soot content, correspondingly.

The integral in Formula 4.11 can be divided into two parts:

(1) for angle range from $\theta_{1cr} = arcsin(n_{2min}/n_1)$ to maximal aperture angle in the light guide θ_{1max}, where condition of TIR is satisfied we can write the following expression neglecting the losses in material of the light guide:

$$P_{out}(\theta_1, \bar{n}_2) = P_{in}(\theta_1) \cdot R_{TIR}(\theta_1, \bar{n}_{2\min}),$$

where $R_{TIR}(\theta_1, \bar{n}_2)$ is calculated from Formulas (4.8) and (4.9).

(2) in the angle range from θ_{1min} till θ_{1cr} TIR condition is not satisfied and therefore,

$$P_{out}(\theta_1, \bar{n}_2) = P_{in}(\theta_1) \cdot R_S(\theta_1, \bar{n}_{2\min}),$$

where $R_s(\theta_1, \bar{n}_2) = 0,5[R^m_{\perp n}(\theta_1, \bar{n}_2) + R^m_{//n}(\theta_1, \bar{n}_2)]$ (see Eqs. 4.5 and 4.6).

Consequently, Eq. (4.11) takes the form:

$$P_{\max} = \int\limits_{\theta_{1\min}}^{\theta_{1cr}} P_{in}(\theta_1) \cdot R_s(\theta_1, \bar{n}_{2\min}) d\theta_1 + \int\limits_{\theta_{1cr}}^{\theta_{1\max}} P_{in}(\theta_1) \cdot R_{TIR}(\theta_1, \bar{n}_{2\min}) d\theta_1. \quad (4.12)$$

Minimal power is defined similarly:

$$P_{\min} = \int\limits_{\theta_{1\min}}^{\theta_{1cr\Delta}} P_{in}(\theta_1) \cdot R_s(\theta_1, \bar{n}_{2\max}) d\theta_1 + \int\limits_{\theta_{1cr\Delta}}^{\theta_{1\max}} P_{in}(\theta_1) \cdot R_{TIR}(\theta_1, \bar{n}_{2\max}) d\theta_1, \quad (4.13)$$

where $\theta_{1cr\Delta} = \arcsin\left(\frac{n_{2\max}}{n_1}\right)$.

Obviously if the number of reflections is large enough then we can assume $R_s(\theta_1, \bar{n}_2) = 0$. In this case the modulation index according to equation will have the following form:

$$M = 1 - \frac{\int_{\theta_{1cr\Delta}}^{\theta_{1\max}} P_{in}(\theta_1) \cdot R_{TIR}(\theta_1, \bar{n}_{2\max}) d\theta_1}{\int_{\theta_{1cr}}^{\theta_{1\max}} P_{in}(\theta_1) \cdot R_{TIR}(\theta_1, \bar{n}_{2\min}) d\theta_1}. \quad (4.14)$$

The modulation index M depends on relative position of RS aperture and LG. At wide RS diagram, when the beam divergence angle exceeds the difference $\theta_{1cr\Delta} - \theta_{1cr}$, such arrangement of RS and LG should be applied when $\theta_{1\max} = \theta_{1cr\Delta}$. However, at such arrangement the maximal output power is decreased. It is possible to avoid this disadvantage with more powerful RS or by narrowing RS diagram.

When the RS aperture is wider, the linear measurement range of refractive indexes increases but difference between maximal and minimal output signal diminishes in this range.

If the test medium has a constant real part of the refractive index ($n_{2max} = n_{2min}$ and $\theta_{1cr} = \theta_{1cr\Delta}$) then formula (4.14) can be written as follows:

$$M = 1 - \int\limits_{\theta_{1cr}}^{\theta_{1\max}} \frac{R_{TIR}(\theta_1, \bar{n}_{2\max}) d\theta_1}{R_{TIR}(\theta_1, \bar{n}_{2\min}) d\theta_1}.$$

Linear range of measured refractive indexes depends as well on the refractive index of the light guide material. Maximal test refractive index should not be higher than refractive index of the LG material.

Calculation of parameters of the light guide soot detectors

In practice the light guide can be made as the optical parallel-sided plate or as the optical cylindrical rod. Analysis of radiation propagation in the light guide presented above is correct for both plate and cylinder (rod) light guides. In common case the technique provides the execution of the following main steps—choice of radiation source, calculation of parameters of the light guide and evaluation of radiation power at light guide outlet and modulation index.

Techniques of calculations of parameters of the soot detectors with plate and cylindrical light guides are presented below.

Calculation of parameters of the soot detector with plate light guide

(1) *Choice of radiation source.* RS should have minimal dimensions, high enough power and its aperture in air should be closed to

$$\varphi_{ap} = \arcsin(n_0 \sin(\theta_{1cr\Delta} - \theta_{1cr})/n_1). \tag{4.15}$$

Radiation spectrum should correspond to the wavelength on which refractive index undergoes maximal changes.

(2) *Calculation of parameters of a plate LG.*

At large thickness of the flat guide the plate ends beveling at angle α_0 for inlet and outlet of radiation in the guide is more advantageous structurally than application of prisms (Fig. 4.19).

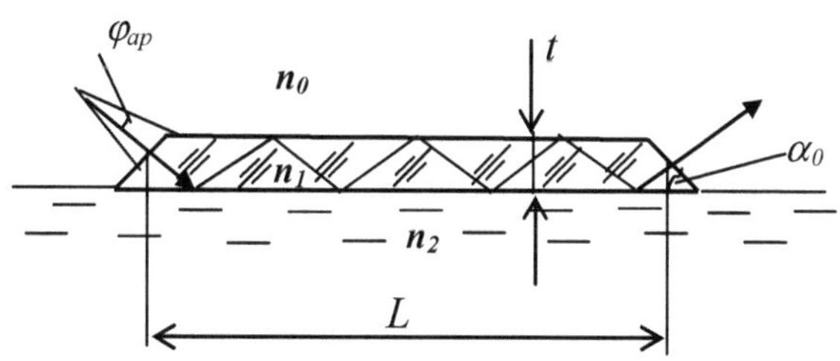

Fig. 4.19 Parameters of a plate light guide

- The refractive index of LG material is estimated. There is an empirical formula for determination of the refractive index of light guide proceeded from the given maximal value of the examined refractive index:

$$n_1 = K(\varphi_{ap}) n_{2\max},$$

where $K(\varphi_{ap})$ is a coefficient depending on aperture width of the source.

The relationship between $K(\varphi_{ap})$ and aperture width is approximated with the error less than 2.5% by the expression

$$K(\phi_{ap}) = 1 + b/\phi_{ap},$$

where b is a constant depending on light guide shape (value of c is in the range from 3 to 5° for the plate guide).

Thus, refractive index n_2 is estimated based on the formula

$$n_1 = (1 + b/\phi_{ap}) n_{2\max}. \tag{4.16}$$

- The angle of radiation propagation in the plate is calculated by the expression $\theta_1 = \arcsin(n_{2\min}/n_1)$. RS diagram should be taken into account in case of the non-collimated beam.
- The beveling angle α_0 at plate ends is chosen according to RS diagram. At narrow diagram it is necessary to be guided by the formula:

$$\alpha_0 = \theta_1 - \arcsin[n_0/n_1) \sin \phi_{ap}], \tag{4.17}$$

where n_0—refractive index of the medium (air) where RS is located.

At wide diagram it is obviously expedient to accept

$$\alpha_0 = \theta_1, \tag{4.18}$$

since number of reflections on the given length decreases sharply at large α_0 angles.

(3) *Choice of the photodetector*. Basic requirement to photodetector selection is a correspondence of spectral sensitivity of the photodetector to spectrum of RS. Besides, size of the sensitive area should be enough to receive the full beam of radiation without vignetting.
(4) *Evaluation of radiation power at light guide outlet and modulation index.* Evaluation of radiation power at light guide outlet is based on Formulas (4.12) and (4.13).

Modulation index can be defined by Eq. (4.14). The number of reflections can be calculated from equation following from Fig. 4.19.

$$m = \frac{L}{2 \cdot t \cdot tg\alpha_0} + \frac{1}{2}.$$

Development of the plate light guide is presented below.

Depending on the type of machine (diesel engine vehicle or locomotive), i.e. regimes of loading, the particles of soot have different parameters (size, composition, and concentration). Therefore, probably, change of the oil refractive index and absorption conditioned by the presence of soot particles will have a maximum on different wavelengths. Besides, limiting content of soot is, obviously, different, i.e. the range of refractive index variation (n_{2max}–n_{2min}) is different for different machines. For instance, according to laboratory measurement the refractive index of the locomotive engine oil varies from 1.49 (fresh oil) till 1.63(used oil). In case of diesel vehicle engine oil the refractive index varies from 1.48 till 1.51 at 940 nm wavelength. Consequently, operation wavelength and range of refractive index variation should be established as initial data for evaluation of detector parameters.

The detector parameter calculation have been performed for condition when refractive index varies from $n_{2min} = 1.47$ till $n_{2max} = 1.59$ for two cases—(1) when RS aperture is wide and (2) when RS aperture is narrow.

RS use with a wide aperture, for example, optical polymer fiber end with light beam diameter *1 mm* and beam divergence angle (half angle) $\varphi_{ap} = 28°$. Light emitter diode (LED) with red wavelength or infrared diode can be used at inlet depending on parameters of soot particles.

The refractive index of the light guide is chosen according to Formula (4.16) $n_1 = (1 + 4/28) \cdot 1.59 = 1.81$. From the nomenclature of optical materials produced in Belarus and Russia the glasses TF10 ($n_1 = 1.8138$) and TBF25 ($n_1 = 1.8175$) which have refractive indexes close to calculated one can be selected.

In this case (wide aperture), the beam divergence angle θ_1 and angle at plate ends α_0 are equal to 54.5° (Formula 4.18).

Photodiode SP–45ML with a large sensitive area (∅7.6 mm) and wide spectral range (λ = 450–1050 nm) can be used as a photodetector.

A disadvantage of this type of detector is low effectiveness (low modulation index) because not full RS radiation is modulated at variation of refractive index within its full range.

RS use with a narrow aperture is more effective.

The optimal RS for a plate LG made of the material with refractive index ~1.81 at given variation of oil refractive index is RS with aperture

$$\phi_{ap} = \arcsin\left\{\frac{n_1}{n_0}\sin\left(\frac{\theta_{1cr\Delta} - \theta_{1cr}}{2}\right)\right\} = \arcsin\left\{\frac{1,81}{1}\sin\left(\frac{61,5 - 4,5}{2}\right)\right\} \approx 6,5°.$$

In the case the modulation index is close to 100% at the plate ends angle calculated by

$$\alpha_0 = \theta_{1cr} + \arcsin[(n_0/n_1)\sin\phi_{ap}] \approx 60^0$$

Probably, we can apply IR-diode EL-1KL3 (light diameter is *4.55* mm, $P_{RS} = 200\ mW$, $\varphi_{ap} = 8°$) too. But in this case a diaphragm should be used to decrease the diameter of the light beam that decreases radiation power.

In the course of RS (its aperture) it is reasonable to use a definition "critical angle" (φ_{cr}) as the light angle in air calculated from critical light angle in light guide θ_{cr} by formula:

$$\phi_{cr} = \arcsin\left[\frac{n_0}{n_1}\sin(\theta_{1cr} - \alpha_0)\right].$$

In Fig. 4.20 relations between refractive index of test oil and critical angle in air (φ_{cr}) and in optical material (θ_{1cr}) are presented for $\alpha_0 = 60°$. Figure 4.20 shows that if RS with diagram $\varphi_{ap} = \pm 8°$ is employed then detector is sensitive to variation in oil refractive index from 1.48 till 1.63.

Thickness and length of the plate light guide are defined taking into account the diameter of the inlet radiation beam, design considerations and manufacturing potentialities (limiting thickness—length ratio for plate element is 1:20). If thickness t is equal to 3 mm and length is about 36.4 mm then the number of reflections can be found as $m = \frac{L}{2 \cdot t \cdot tg\alpha_0} + \frac{1}{2} = \frac{36.4}{2 \cdot 3 \cdot tg60^0} + \frac{1}{2} = 4.$

Basic feasible designs of the optical plate sensor and built-in oil monitor are presented in Fig. 4.21. Optical radiation could input in the light guide from RS and output to PD directly (Fig. 4.21a, b) or with transmission radiation by means of the optical fiber (Fig. 4.21c). The light guide element can have the reflective surface

Fig. 4.20 Dependencies of critical angle in air (φ_{cr}) and in optical material (θ_{1cr}) on test oil refractive index n_2

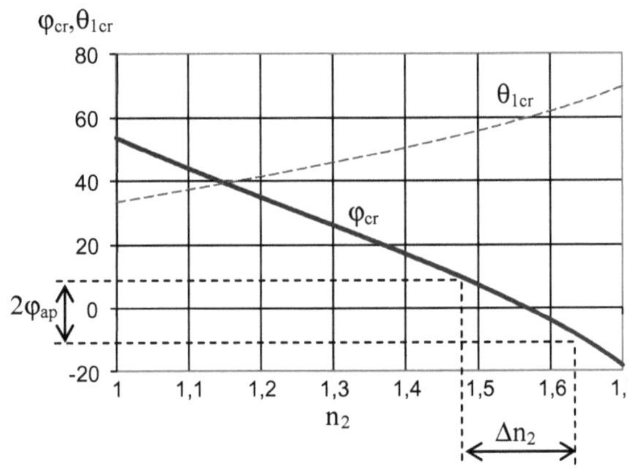

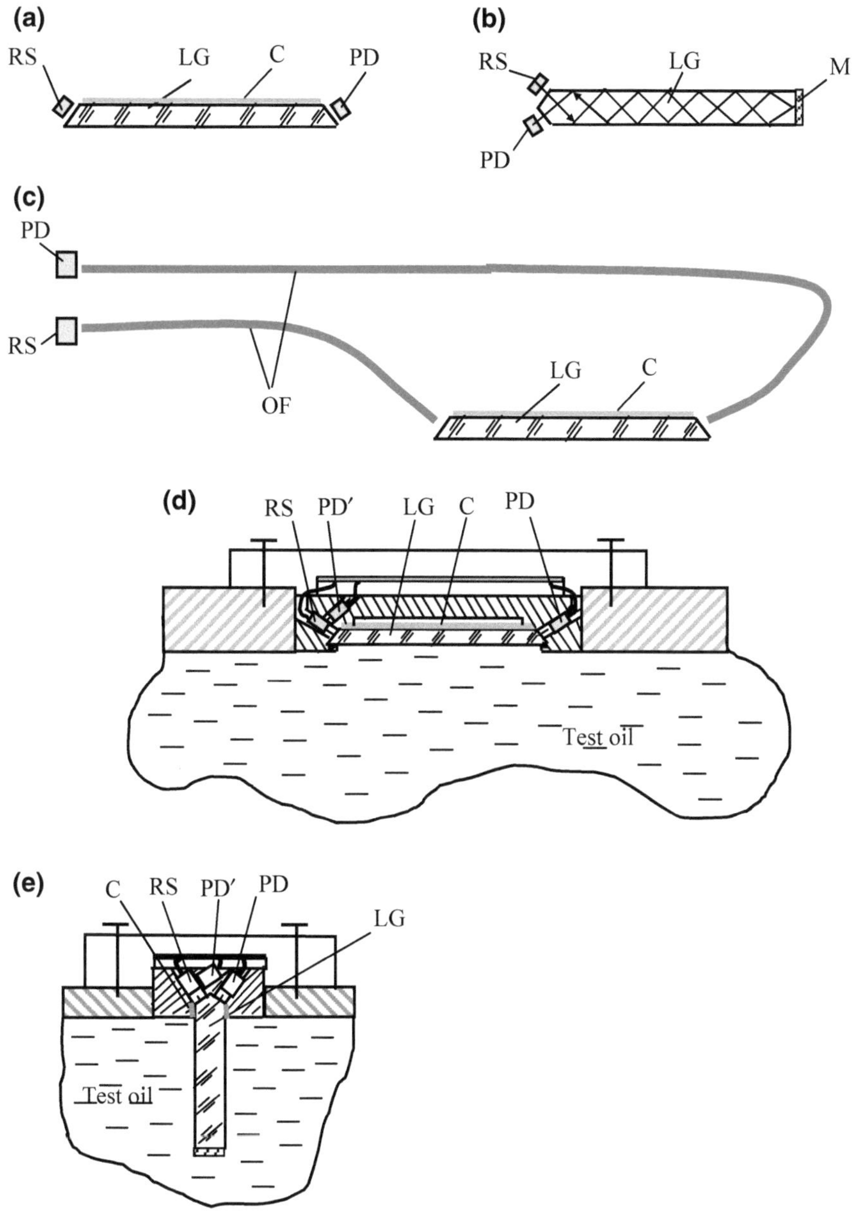

Fig. 4.21 Feasible designs (**a**, **b**, **c**) of optical plate sensor; the oil detector built in oil flow line (**d**) and detector built in test oil tank (**e**): LG, RS, PD, PD′, M, OF and C are light guide element, radiation source, photodetector, reference photodetector, mirror, optical fiber, and coating, correspondingly

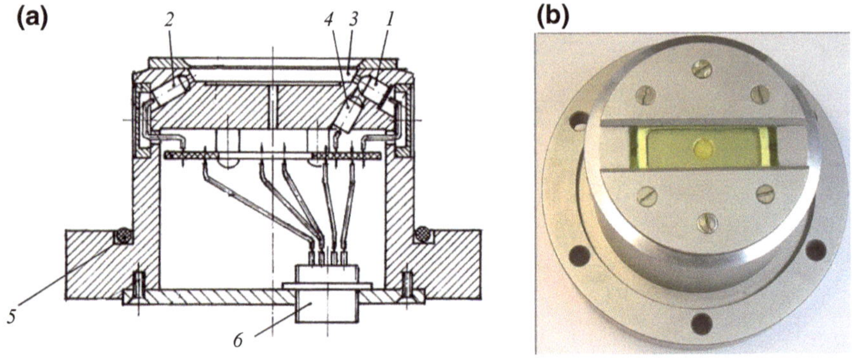

Fig. 4.22 Soot detector with plate light guide: design (a): *1* radiation source; *2*, *4* photodiodes; *3* optical plate; *5* gasket; *6* electrical connector; photo (b)

(3) in order to simplify design and to decrease sensor dimensions. The light guide element surface which is not bound with the test oil has protective coating (C) with the refractive index lower than the refractive index of the light guide element.

The detectors with the optical plate sensor can be used in the oil flow line (Fig. 4.21d) and in the test oil tank (Fig. 4.21e).

The soot detector manufactured according to mentioned design is presented in Fig. 4.22.

A sensitive element of the sensor is the optical light guide plate 3, made of the optical glass TF-10. The bevel angles (α_0) of plate ends are 60° to input and output the optical radiation into the plate.

IR–diode EL-1KL3 with wavelength λ = 940 нм is used as a radiation source 1. Optical light passed through the plate is converted into the electrical signal by means of the photodiode PD–256. Feed back with the photodiode 4 output as input signal is applied to stabilize IR-diode radiation. The detector is installed in oil tank with the gaskets 5. The mechanical unit is conjugated through the connector 6 with electrical unit to proceed and indicate information on soot content in oil.

The detector allows us to monitor the soot content in oil in range from 0.5 to 10.0 wt%.

Decision on parameters of detector with the cylindrical light guide (optical rod) for monitoring the diesel engine oils

Optical rod and test oil can be considered as an "optical fiber" in which the core is an optical rod and cover is the test oil. To realize condition of TIR the refractive index of the rod (n_1) should be greater or equal to refractive index of the test oil (n_2).

The relation between the critical incident angle in the light guide (θ_{1cr}) and in the air ($\varphi_{cr,}$) is (Fig. 4.23)

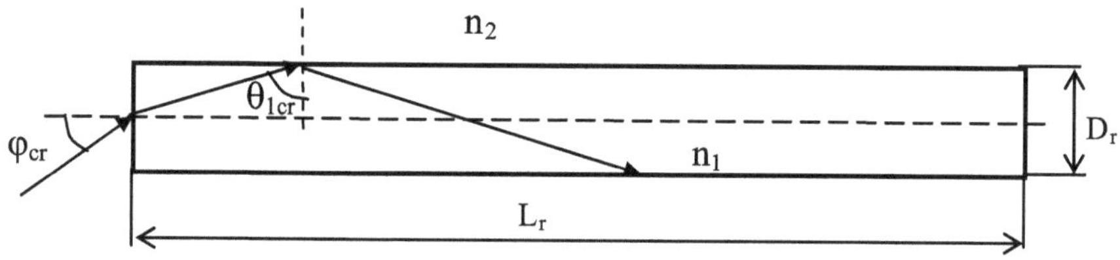

Fig. 4.23 The optical rod as a light guide

$$\sin(\frac{\pi}{2} - \theta_{1cr}) = \cos\theta_{1cr} = \frac{n_a \sin\phi_{cr}}{n_1} = \frac{1 \cdot \sin\phi_{cr}}{n_1}.$$

Taking into account the formula

$$\theta_{1cr} = arcsin\frac{n_2}{n_1} \Rightarrow \cos\,\theta_{1cr} = \sqrt{1 - \frac{n_2^2}{n_1^2}},$$

the critical angle in air can be expressed through the refractive indexes of the examined medium (oil) and the light guide as follows:

$$\frac{\sin\,\varphi_{\rm cr}}{\rm n_1} = \sqrt{1 - \frac{n_2^2}{n_1^2}} \Rightarrow \varphi_{cr} = arcsin\sqrt{n_1^2 - n_2^2}$$

It is obvious that refractive index of the optical rod should be selected on the basis of the following requirements:

$$\begin{cases} n_1 > n_2 \\ n_1^2 - n_2^2 \leq 1. \end{cases}$$

If refractive indexes of media are equal to each other, the output power passed through the rod due to TIR is equal to zero. At a minimal value of oil refractive index n_{2min} the RS ray beam satisfied to TIR conditions is limited by the following requirement:

$$\phi_{ap} \leq \arcsin\sqrt{n_1^2 - n_{2\min}^2}.$$

From this expression we can estimate n_1 using given n_{2min} and RS aperture. For example, if n_{2min} = *1.47* and $\varphi_{\rm ap} = 28°$ then the refractive index of light guide should be defined as follows:

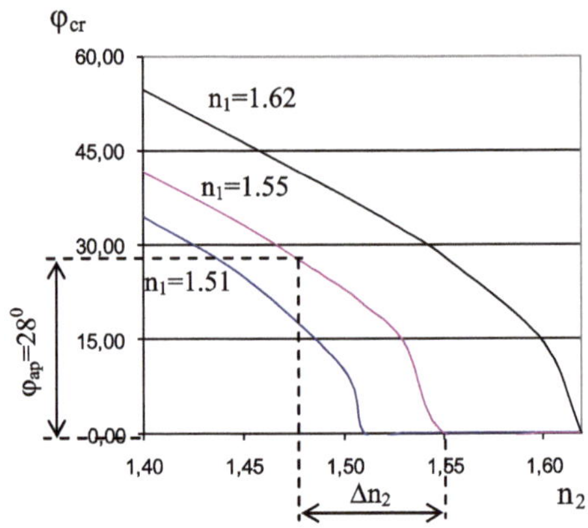

Fig. 4.24 Dependencies on oil refractive index n_2 of the critical angle in air φ_{cr} for different material (n_1 = *1.51, 1.55 and 1.62*) of optical rod

$$n_1 \geq \sqrt{\sin^2(\phi_{ap}) - n_2^2} = \sqrt{\sin^2(28^0) + 1,47^2} \cong 1.545.$$

That is, the full beam of the radiation will be involved in measurement of n_2 (n_{2max} = 1.55) at n_1 = 1.55 (Fig. 4.24).

The average number of reflections is evaluated by formula:

$$m = \frac{L_r \cdot ctg\theta_{1cr,\min}}{D_r}.$$

If L_r = *80 mm*, D_r = *3 mm*, n_1 = *1.55 and* n_{2min} = *1.47* then m = *4*.

Radiation power at the rod outlet can be defined by the Eqs. (4.12) and (4.13).

In the case when the real part of refractive index of the test oil does not vary and just the absorption index (imaginary part) varies the Eq. (4.14) for modulation index calculation takes the following form:

$$M = 1 - \frac{\int_{\theta_{1cr}}^{\theta_{1\max}} R_{TIR}(\theta, \chi_{\max}) d\theta_1}{\int_{\theta_{1cr}}^{\theta_{1\max}} R_{TIR}(\theta, \chi_{\min}) d\theta_1}. \tag{4.19}$$

Let us estimate the absorption index χ of the used diesel oil.

The absorption index χ has the following relation with extinction coefficient in case of soot analysis, that is when scattering is considerably less than absorption

$$\chi = \frac{\lambda}{4\pi} K_{ext,\lambda}$$

If λ = *670 nm* and, c = *1%* then χ = *0.002* taking into account the equality $\alpha_\lambda = k_\lambda c$, where $k_\lambda \approx 40 \frac{1}{\text{mm}}$ for oil at $\lambda \approx 700$ nm.

For rough evaluation of modulation index the values of integrals in Eq. (4.19) can be taken as squares under correspondent curves $R_m(\theta_1)$ in Fig. 4.15. In the case $n_1 = 1.51$, $n_2 = 1.48$, $m = 4$ and $\chi = 0.002$ the modulation index is $M \cong 1 - \frac{0.75}{1} = 0.25$ (25%).

Design of the soot detector with the optical rod

Basic feasible designs of the soot detector with the optical rod sensor are presented in Fig. 4.25 [31].

Optical radiation can be put to the optical rod from RS and output from the optical rod to PD directly (Fig. 4.25a, b) or through the optical fibers (Fig. 4.25c). The light guide element can have the reflective surface (3) in order to simplify the design and to decrease the sensor dimensions. The light guide element has a coating (C) with refractive index lower than refractive index of light guide element in zones, which are not bound with test oil for protection of light guide surface from soiling (dirtying) and damage of the rod.

The monitor with optical rod sensor can be used in the test oil tank (Fig. 4.25d) and in the oil circulation line (Fig. 4.25e).

In Fig. 4.26 the soot detectors based on the mentioned designs are shown.

In case of the on-line detector the optical rod made of glass BK8 and bush 2 with the inlet and outlet optical fibers are mounted in the housing 3. The inlet fiber transmits the radiation from RS (IR—diode EL–1ML2) to optical rod. The radiation reflected from mirror coating passes to PD (SP–1ML) through the outlet fiber. To apply an electrical field (see below) the electrode 4 is installed in alignment with the optical rod. Test oil is pumped through the passage 6.

Laboratory test of the soot detector with optical rod

The main purpose of laboratory test is to verify theoretical estimation of parameters of optical rod sensor presented above.

Optical rods made of glasses K8 (n_D = 1.51) and BK8 (n_D = 1.55) were tested. LED of wavelength 670 nm and IR—diode LED of wavelength 940 nm were used as optical radiation sources.

Test procedure. Oil samples with 0, 1, 2, 3 and 5 wt% soot concentration were prepared using clean diesel engine oil “API CF-4” with viscosity 15 W-40 and carbon black powder “Color Black S 170, Type MCC”. In order to make dispersed enough colloidal oil sample the homogenizer was used during about eight hours.

The real used oil “API CF-4” draining from the diesel truck was tested.

The tests were carried out at temperature 21–22 °C. The test procedure included the following steps:

- sensor is in air and output signal is written during 100–200 s;
- sensor is dipped into a bottle with oil sample and output signal is written during 800–900 s;
- sensor is washed in ultrasonic bath with hexane.

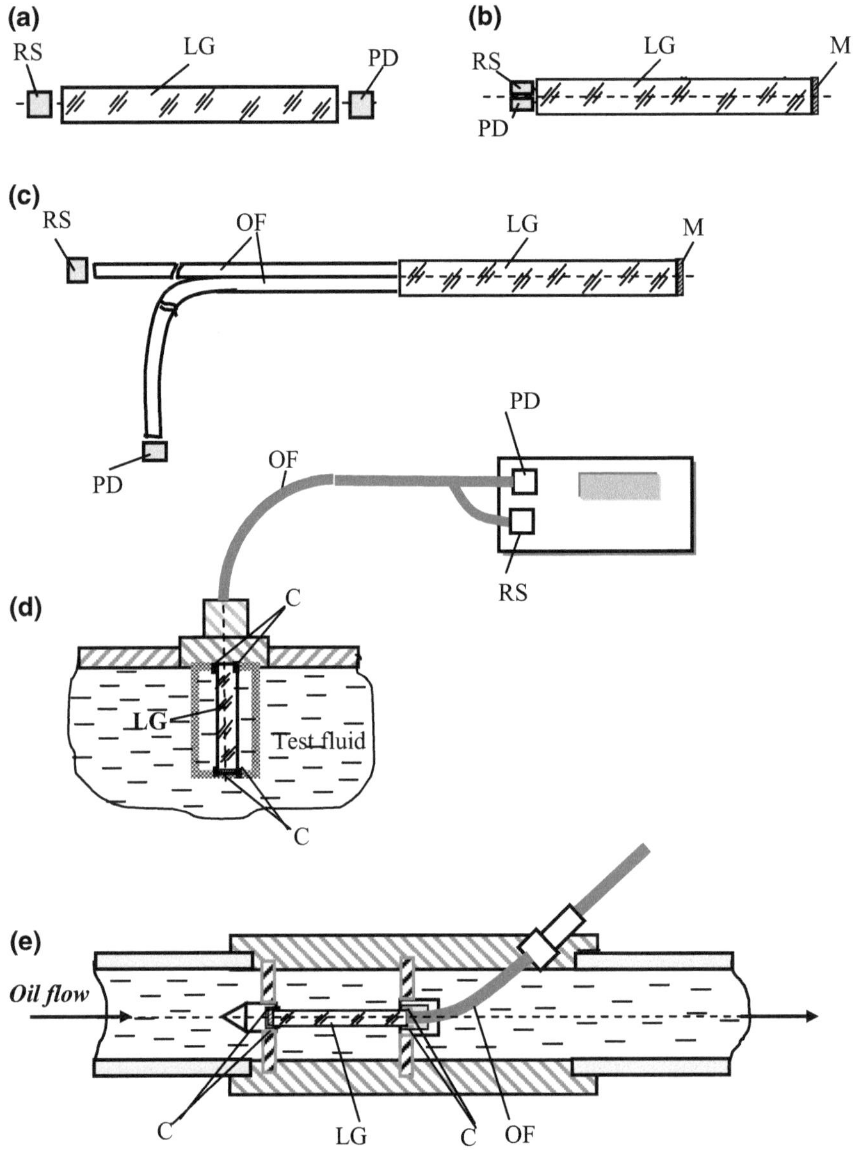

Fig. 4.25 Feasible designs (**a**, **b**, **c**) of optical rod sensor and built-in oil monitor in test oil tank (**d**) and in oil flow line (**e**): LG, RS, PD, M, OF and C are light guide element, radiation source, photodetector, mirror, optical fiber and coating, correspondingly

Figure 4.27 presents the dependences of percentage change in the output signals on carbon black concentration for the rod materials K8 and BK8 at different wavelength of radiation source (λ = 670 and 940 nm). Results of the used oil test are shown in the figure.

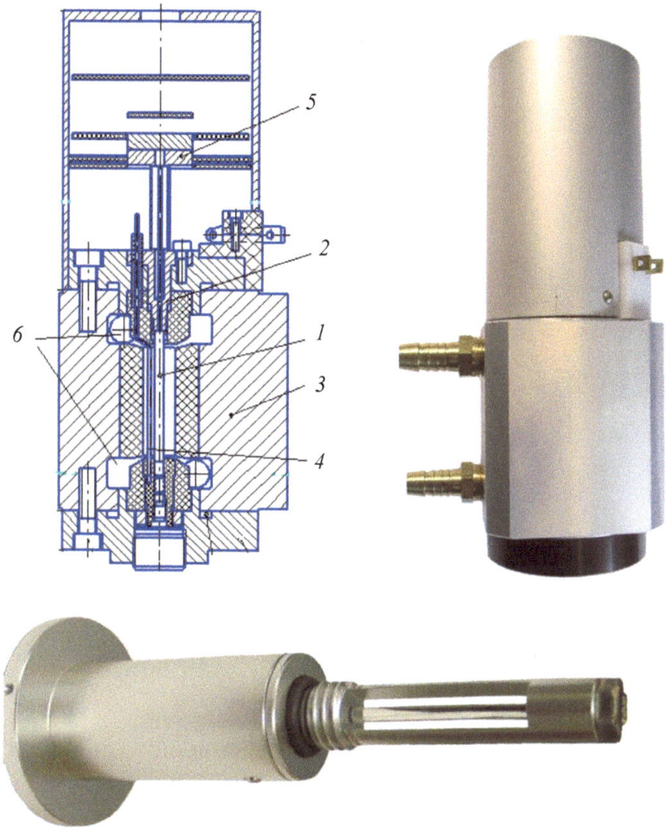

Fig. 4.26 Soot detectors with the optical rod: **a** online type; **b** in-line type. 1 optical rod, 2 bush with inlet and outlet optical fibers, 3 housing, 4 electrode, 5 bush with RS and PD, 6 oil passage

Clearly, that when the radiation wavelength increases then sensitivity of the detector rises and percentage change in output increases. It is caused by the increase of light penetration depth into the oil when the wavelength increases.

Thus, we can change the measured range of concentration choosing the type of rod material and RS wavelength. If we need to monitor the soot concentration in the range from 0 to 5 wt% then the rod made of material with refractive index $n_D = 1.55$ and radiation source with wavelength $\lambda = 940$ nm are optimal. To provide the measurement range from 0 to 10 wt% the rod with $n_D = 1.51$ and radiation source with $\lambda = 670$ nm are preferable.

Solution of contamination problem of optical sensitive surface of light guide

The main problem occurred in the course of soot detector development is the high adsorption of soot particles on the surface of optical rod. As a result the fast surface contamination and abnormal performance of the detector occur.

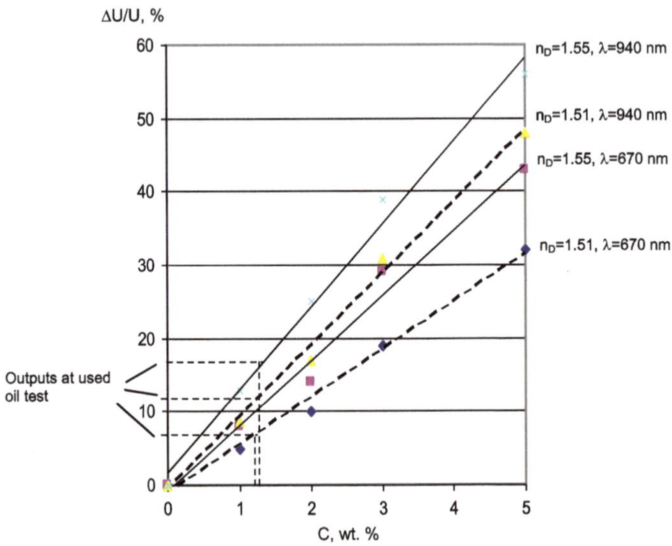

Fig. 4.27 Concentration dependence of output signal change for rod materials with refractive indexes n_D = 1.51 and 1.55 and source wavelength 670 and 940 nm

Three aspects should be taken into account when analyzing the interaction between soot particles and optical surface of sensor:

- stability of the colloidal system of soot particles in oil;
- soot particle adsorption on the surface;
- particles adhesion to the optical surface.

Stability of the colloidal system determines the size of particle aggregates that affects adsorption and interaction intensity between particle and surface.

Stability of the colloidal system "soot particles in oil"

Diesel used oil can be considered as a colloidal system because of the small size of soot particles (20–40 nm). In order to keep soot particles suspended in oil we have used the dispersive additives which are surface-active substances.

Coagulation of the soot particles happens due to Van der Waals attraction forces between the particles at sufficient approach. The attraction forces between the dispersed particles are induced by the attraction of their electrical dipoles (so-called disperse forces). F. London has calculated the attraction between two electrical point dipoles (E_v) and has shown that attractive energy decreases proportionally to l_e^{-6} and H.C. Hamaker has derived an equation for the attractive energy between two spheres having the same diameter D [12].

$$E_v = -\frac{H}{6}\left(\frac{2}{l_e^2+4l_e} + \frac{2}{(l_e+2)^2} + \ln\frac{l_e^2+4l_e}{(l_e+2)^2}\right), \qquad (4.20)$$

where $l_e = \frac{2s}{D}$, s is a distance between the surfaces of spheres, H—Hamaker's constant which is dependent on dielectric properties of particles and medium and determines value of disperse forces in a given system.

According to Eq. (4.20) the attractive energy of solid spheres in contact ($l_e \to 0$) becomes infinitely large and coagulation of particles takes place. When the distance between particles increases the attractive energy decreases proportionally to l_e^{-6} (similar to London's theory).

Aggregative stability of a colloidal system is protected by surface adsorptive layers preventing surface approach to a distance at which energy of attraction will be higher than the disordering effect of thermal motion. To keep the soot particles in suspended condition in oil (stable colloidal system) the complex of detergent additives is injected in oil. Main function of the detergent additive is production of an adsorptive layer preventing the attraction between the particles.

Thickness of the adsorption layer δ should be enough to follow the condition

$$|E_v| \leq kT. \tag{4.21}$$

It means that disordering effect of the thermal motion prevents particles from adhesion.

Boundary $|E_v|/kT = 1$ separates conditions of aggregative stability and instability of the fluid, where the dispersed phase coagulates when solid particles collide.

Let's analyze the dependence of E_v on particle size. According to relation (4.20) E_v is a function of δ/D ratio at the contact of adsorptive layers.

Above arguments relate to stabilized particles, distance between which is not less than 2δ. However, in the course of Brownian collision of particles the deformation of the adsorptive layer can occur and as a result particles are able to approach to a distance s < 2δ. In this case, the mechanism of steric repulsion of particles protects stabilized particles from further approach. This mechanism is caused by interaction of adsorptive layers of long-chain molecules at close approach of particles.

Energy of the steric repulsion can be estimated using the formula [12]

$$\begin{aligned} \frac{E_s}{kT} &= 2\pi D^2 N\left(2 - \frac{l_e + 2}{t_e}\ln\left(\frac{1 + t_e}{1 + (l_e/2)}\right) - \frac{l_e}{t_e}\right), \quad \text{at}\, l_e < 2t_e \\ \frac{E_s}{kT} &= 0, \quad \text{at}\, l > 2t \end{aligned} \tag{4.22}$$

where $t_e = 2\delta/D$; N is the concentration of adsorbed molecules on the surface.

Electrostatic repulsion arising from the interaction of ions adsorbed on the particle surface is neglected because of low ion concentration and low dielectric constant of the fluid medium. The result of algebraic addition of energies of Van der Waals attraction and steric repulsion for particles with diameters D = 20 and 40 nm is presented in Fig. 4.28.

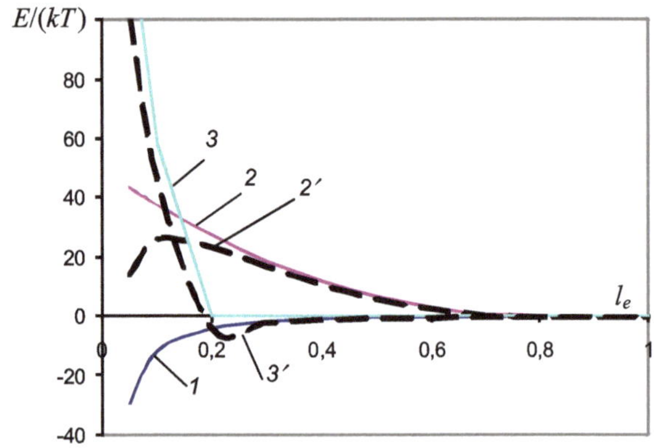

Fig. 4.28 The dependence of potential energy on distance *l* between surfaces of colloidal particles with steric coating; 1 Van der Waals attraction; 2, 3 steric repulsion of particles with 10 and 40 nm diameters, correspondingly; 2′, 3′ resulting interaction of particles with diameters 10 and 40 nm, correspondingly

In calculations of steric repulsion it was assumed that 10^{18} molecules of surfactant adsorbed per square meter of the solid particle surface. Hamaker's constant was taken as $H = 10^{-19}$ Nm at calculation of Van der Waals attraction.

When concentration and size of dispersed particles increase, conditions are existed for forming aggregates in oil due to appearance of minimum of resulting potential energy at enough large distances (Fig. 4.28, curve 3′).

Distribution of solid particles is conditioned by diffusion processes in the fluid in thermodynamic equilibrium state.

Adsorption of soot particles on optical surface

Adsorption is a process of spontaneous redistribution of system components between the surface layer and volume phase. The adsorption is caused by the tendency to spontaneous decreasing of the surface energy which means decreasing of surface tension.

Two quantities are applied in order to describe adsorption. One of them (*A*) is the number of moles or grams of adsorbate per unit surface or per unit mass of the adsorbent. Another characteristic of adsorption is determined by the excess of substance (adsorbate) in the surface layer as compared to its quantity in the same phase volume related to the unit square surface or to unit mass of the adsorbent. This quality is called Gibbs adsorption (*G*).

When equilibrium in the system is reached, the quantity of adsorbed substance by the surface layer depends on the temperature and concentration of this substance in phase volume.

The equation of the system state in relation to adsorption value *A* has the form:

$$A = f(c_a, T), \tag{4.23}$$

where c_a is the equilibrium concentration of the adsorbate, T is temperature.

The differential relation is widely used in thermodynamics

$$\left(\frac{\partial A}{\partial c_a}\right)_{T=const} \cdot \left(\frac{\partial c_a}{\partial T}\right)_{A=const} \cdot \left(\frac{\partial T}{\partial A}\right)_{c_a=const} = -1. \tag{4.24}$$

Derivative $(\partial A/\partial c_a)$ is always positive because chemical potential and, accordingly, activity of the adsorbate becomes uniform across the whole system at equilibrium state. Therefore, when concentration (pressure) of the adsorbate increases in the bulk its concentration in surface layer, that is, value of adsorption A, rises. Thus two other derivatives have different signs.

The equation of adsorption system state can be also written relatively to the surface tension as follows:

$$\sigma = \phi(c_a, T) \tag{4.25}$$

Then, differential relation will have the form

$$\left(\frac{\partial \sigma}{\partial c_a}\right)_{T=const} \cdot \left(\frac{\partial c_a}{\partial T}\right)_{\sigma=const} \cdot \left(\frac{\partial T}{\partial \sigma}\right)_{c=const} = -1. \tag{4.26}$$

The relation between the surface tension and chemical potentials μ_i of system components is described by a fundamental Gibbs adsorption equation:

$$-d\sigma = \sum_i G_i d\mu_i, \tag{4.27}$$

where G_i is the value of Gibbs adsorption. Excess of adsorbate on the surface was related by Gibbs to a unit surface in assumption that surface layer has no volume. In order to obtain an equation for Gibbs adsorption G_i let us suppose that N_{total} is the total number of particles *ith* substance in the system, c_i is the equilibrium bulk concentration of *ith* substance, V is the phase volume and S_q is the square of interface. Then the value of Gibbs adsorption can be written as follows:

$$G_i = \frac{N_{total} - c_i V}{S_q} = \frac{V(c_{0i} - c_i)}{S_q},$$

where c_{0i} is initial concentration of i substance in the system.

Taking into account thickness h_s and volume V_s of the surface layer (method of "finite thickness layer") Gibbs adsorption can be described as follows:

$$G_i = \frac{V_s(c_b - c_i)}{S_q} = \frac{A_i S_q - c_i V_s}{S_q} = A_i - c_i h_s,$$

where c_b is bulk concentration of *ith* substance in the surface layer and A_i is adsorption of the *ith* substance.

The driving force of adsorption is the difference of chemical potentials of components in the bulk and on the surface, which become even in time. As a result, the adsorption equilibrium sets in. The process of adsorption can be very fast in some cases or rather slow in other cases and sometimes lasts long (a day or more). Adsorption kinetic is changing adsorption value A in time τ_a:

$$A = f(\tau_a), \frac{dA}{d\tau_a} = f'(\tau_a).$$

The derivative describes the adsorption rate. Adsorption rate on homogeneous surface with formation of a monomolecular layer (Langmuir adsorption) is the sum of rates of two opposite processes. The rate of forward process is determined by the rate of attaching adsorbate molecules on the adsorbent surface and the rate of an inverse process (desorption) is the rate of removing molecules from the surface. Rate of adsorption is proportional to the number of collisions of adsorbate molecules with a square unit, free surface percent and the percent of molecules able to fix on the surface. Rate of desorption is proportional to the number of molecules leaving the unit of saturated surface and to the degree Q of surface filling.

Adsorption rate can be written as

$$\frac{dQ}{d\tau_a} = k_a \cdot \left(Q_{eq} - Q\right),$$

where k_a is a constant of adsorption rate which defines the time of reaching adsorption equilibrium.

Hence, it follows:

$$Q = Q_{eq}\left(1 - e^{-k_c \tau_a}\right) \text{ or } A = A_{eq}\left(1 - e^{-k_c \tau_a}\right),$$

where Q_{eq} and A_{eq} are degree of surface filling by adsorbate and adsorption at equilibrium, respectively.

Obtained equation describes the adsorption kinetics presented in Fig. 4.29.

Nevertheless, the fact that not all particles coming near the surface due to diffusion will be fixed on the surface owing to Van der Waals forces should be taken into account. This means that adhesion of particles to surface should be considered.

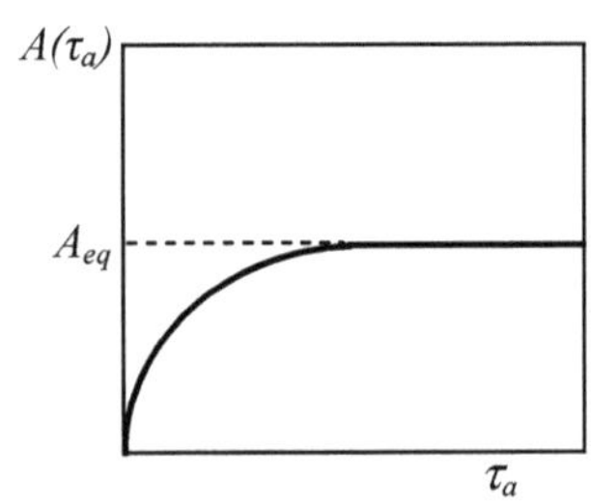

Fig. 4.29 Kinetic curve of adsorption

Adhesion of soot particles to the optical surface

Capillary and electrical forces occurring in gas media are not effective in fluids practically. Consequently molecular interaction is the main contributor in adhesion of contacting bodies. Besides, repulsion forces induced by the fluid arise. Therefore, forces of attraction and repulsion should be considered in detail to define the reasons of adhesion in the fluid medium.

The forces of intermolecular interaction can be evaluated by the Van der Waals constant (Hamaker constant, *H*). Adhesion forces caused by molecular interaction depend on fluid properties and those of contacting surfaces. So, adhesion of graphite particles of 2.5 μm diameter to the quartz (hydrophilic) plate, i.e. contacting surfaces sharply differ from each other in wetting, equals to zero, but to the paraffin surface (particles and the surface are hydrophobic) it reaches rather high value [32].

The effect of capillary, electrical and Coulomb forces in fluid are neglected and adhesion is caused by intermolecular forces (disjoining pressure of the fluid layer between contacting surfaces prevents adhesion). The value of molecular forces is directly proportional to particle size; when one particle is a sphere and another is a plate $F = HR/6h_{sp}^2$, where *R* is radius of the sphere and h_{sp} is the distance between the sphere and plate.

In spite of dispersive additives the soot particles can coagulate forming rather large aggregates. Electron microscope analysis of the soot particles extracted from used diesel oil has shown that initial size of the particles is from 30 nm till 50 nm. In the case of the well dispersed soot in the six-cylinder Mack diesel engine the mean particle size is about 150 nm whereas for the poorly dispersed soot, the mean particle size is about 210 nm [11].

If diameter of the aggregates increases then adhesion of the aggregates to optical surface increases, that is, number of particles kept on the surface rises. Figure 4.30 presents the output of the soot detector dipped in used oil which was not agitated for

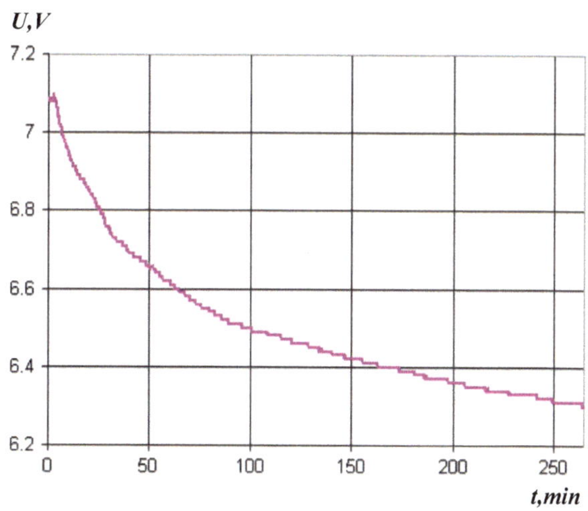

Fig. 4.30 Time dependence of the soot detector output in the course of used oil test at room temperature without excitation

a week. Change of output signal is caused by transition to adsorption equilibrium and by increase of particle adhesion in time.

Oil flow and temperature effects on particle adhesion to the optical surface of the sensor

During air blow of dust horizontal surface the adhesion force F_{ad}, particle weight P, drag F_r and lifting force F_c act on the adhered particle. The condition of particle separation can be presented as follows [32]

$$F_r + F_c \geq \eta(F_{ad} + P), \tag{4.28}$$

where η is the friction coefficient.

For small particles, i.e. when F_{ad} » P and F_{ad} » F_c in equality (4.28) can be simplified to $F_r \geq \mu F_{ad}$. The same conditions (4.28) can be used in case of removing (separating) particles from the solid surface by a liquid flow.

Separating force F_{sep} depends on the density and viscosity of the fluid, the particle diameter, the flow rate and conditions of the flow around the adhered particles.

Obviously, the particle adhesion is affected by the rate of flow circulation, which causes the additional separating force. At high enough flow velocity the desorption rate prevails over the adsorption rate and the output signal of the soot detector increases up to some level corresponding to the equilibrium condition at a given temperature and flow rate. At low velocity the adsorption rate is higher than desorption rate and the output signal decreases down to the equilibrium value. Probably, at high temperature and flow rate the equilibrium condition between adsorption and desorption can be realized when surface and bulk concentrations of soot particles in oil are equal.

Laboratory set-up shown in Fig. 4.31 was used for experimental investigation of external factors effects on the condition of sensitive surface of the light guide [33]. Test diesel engine oil (API CF-4, 15 W-40) with given soot particles ("Color Black S 170, Type MCC") concentrations was put in the oil tank. The oil was pumped with the flow rate adjusted by controller through the oil circulation line where the soot detector was installed. Oil temperature was preset and adjusted by temperature controller. Besides, the oil temperature was measured by the temperature sensor. Outputs of soot detector and temperature sensor were fed to computer through the electronic block.

Effect of the oil flow rate on the soot detector output was tested at temperatures 27 and 50 °C after keeping the detector with the oil with soot content 1 wt% for a week. Obtained results (Fig. 4.32) are the evidence of rising efficiency of surface cleaning at oil flow increasing. Sludge accumulated on the optical rod surface is removed at flow rate of 1 l/min at temperature of 27 °C and at flow rate of 0.8 l/min at temperature of 50 °C. In real diesel engines the oil flow rate exceeds this rate: flow rate is 3.5 l/min in the engine D240 and flow rate is 4.2 l/min in the engine

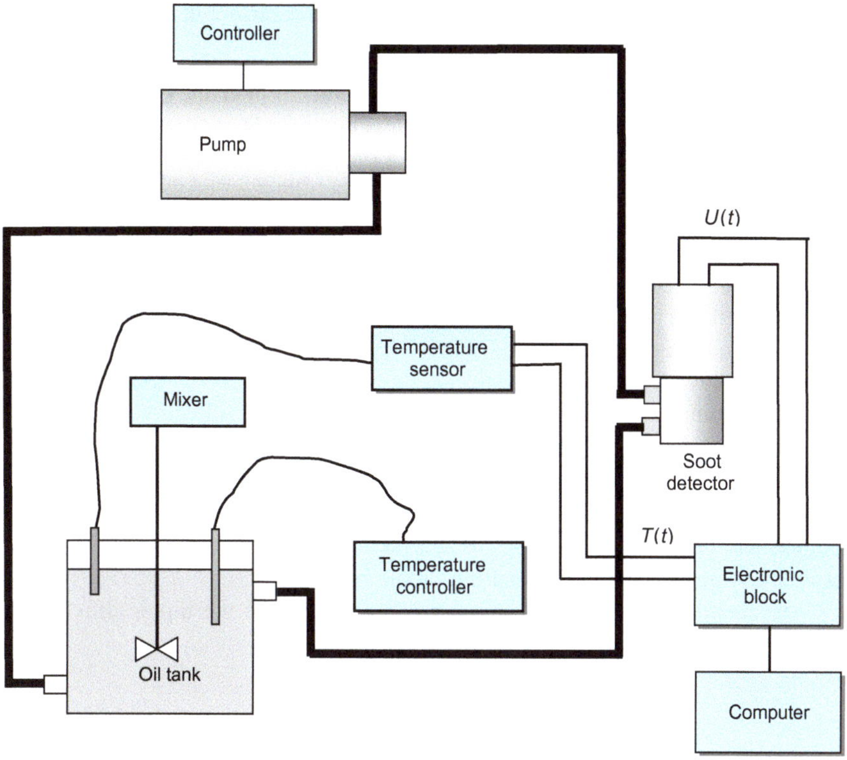

Fig. 4.31 Bench for soot detector test

Fig. 4.32 Dependence of the soot detector output on the oil flow rate in the oil circulation line at temperature of 27 °C (curve 1) and 50 °C (curve 2)

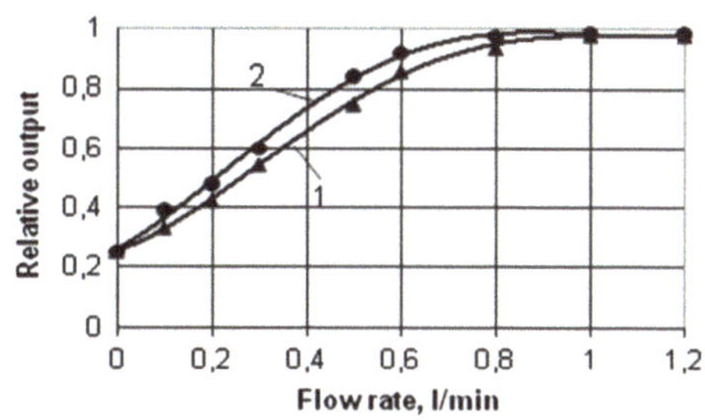

D260. That is, in the course of engine operation the particles desorption from the light guide surface is sufficient to prevent the deposition and accumulation of the particles.

However after long vehicle downtime (for several days) a hard-to-remove soot layer (even at engine start-up) could be formed because of adsorption and adhesion. Obviously, special methods of preventing particle adsorption should be used in this case. Electrical field, ultrasonic radiation and oleophobic coating could be applied.

Analysis of effect of electrostatic field on soot particles adsorbed to optical surface of light guide

When an uncharged conductor is placed in the electrical field, the carriers of charge start to move. Positive carriers move in direction of electric-field vector E and negative supports move in the opposite direction. As a result, induced charges with opposite signs appear on the conductor ends. The direction of the charge field is opposite to the external field.

It seems that the charges are induced on soot particles in the electrical field, so the particle can be considered like a dipole.

Forces acting on the dipolar charge (f_1 and f_2, see Fig. 4.33) are not equal by magnitude in the non-uniform field. Forces f_1 and f_2 can be considered as collinear at small dipole size.

Let us assume that the field gradient is maximal in the direction x coinciding with the direction of the field intensity E in the point, where the dipole is placed. The positive charge of the dipole is shifted relatively to the negative charge by the value $\Delta x = l_d \cdot \cos\alpha$ in x direction. Therefore, difference of the field strength in the points where charges are placed is $\Delta E = \frac{\partial E}{\partial x}\Delta x = \frac{\partial E}{\partial x} l_d \cdot \cos\alpha$, where α is the angle between the gradient direction and the dipole moment.

Consequently, the resultant force $f = f_1 + f_2$ acting on the dipole differs from zero. Projection of the resultant force on axis x equals to

$$f = q \cdot \Delta E = q\frac{\partial E}{\partial x} l_d \cdot \cos\alpha = p\frac{\partial E}{\partial x} \cdot \cos\alpha, \tag{4.29}$$

where $p = ql_d$—the magnitude of the dipole moment.

Thus, in the non-uniform field the dipole experiences a force, which either pulls the dipole into the zone of the stronger field (α is acute angle) or pushes it out of the field (α is obtuse angle).

Let us consider the soot particle behavior in a field of two charged electrodes (conducting cylinders of a small diameter, axes of which are normal to the drawing), (Fig. 4.34).

Charges are induced on the uncharged soot particle at application of the electrical field and the particle obtains dipole properties. Force f_1 aiming to shift the

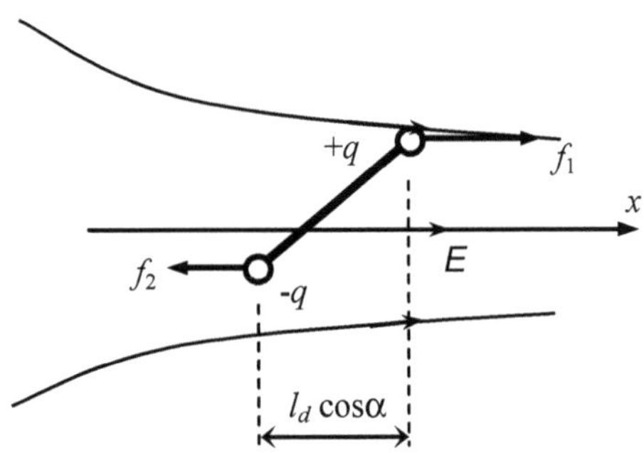

Fig. 4.33 Dipole is in the electrical field E: f_1 and f_2 are the forces acting on positive and negative dipole charges; l_d—dipole length; α is an angle between the "gradient direction" x and the dipole

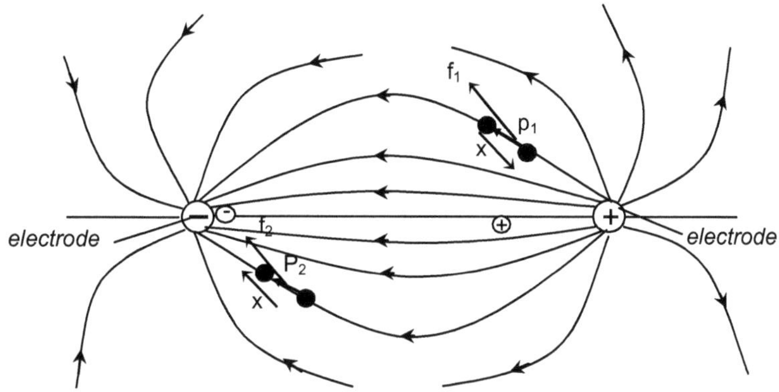

Fig. 4.34 Soot particles in the electrical field of two electrodes

particle into the zone of low field strength, acts on dipole p_1, which is placed near the positive electrode (see Formula 4.29). Force f_2, aiming to shift the particle into the zone of high field strength, acts on dipole p_2, which is placed near the negative electrode. If this force is sufficiently high to overcome the forces of the fluid resistance and oil flow, then the particle will be captured by the negative electrode. In this case, the particle will impart positive charge to the electrode becoming a carrier of negative charge. The negatively charged particle will push out the negative electrode and will be directed towards the positive electrode, where it will change its charge for the opposite and will be directed towards the negative electrode again.

Experimental study of electrostatic field effect on soot particles adhesion to the optical surface was carried out using the soot detector with electrodes which are parallel to axis of optical rod (Fig. 4.35). The detector was installed in the test bench filled by oil with soot concentration of 1 wt%.

Fig. 4.35 Diagram of electrodes arrangement in soot detector (1 is optical rod and 2 are straight electrodes), and scheme of voltage supply to the electrodes: without (**a**) and without (**b**) charge on light guide surface

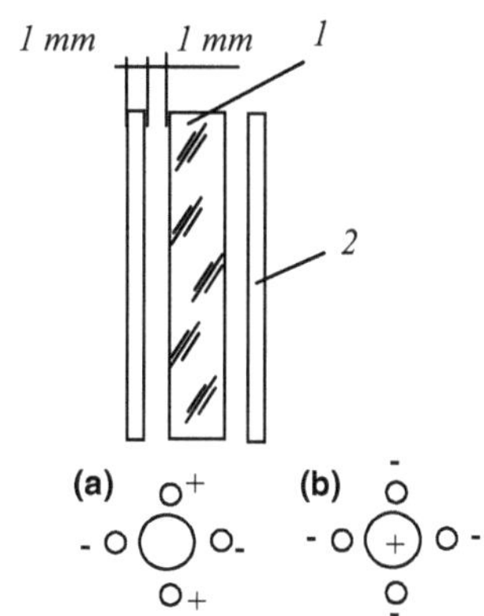

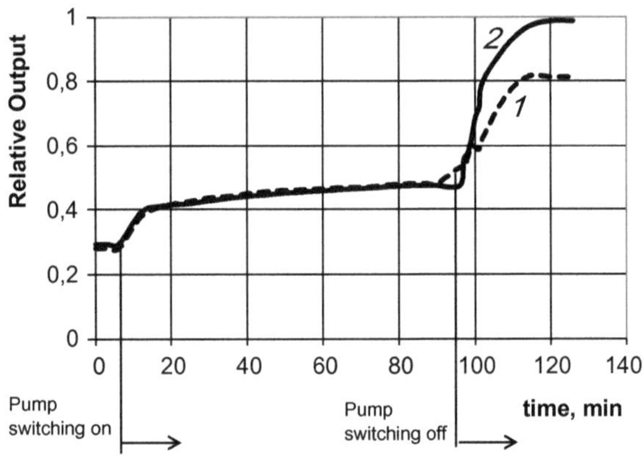

Fig. 4.36 Change of the soot detector output at the electrical field application: 1 light guide without coating; 2 light guide with conducting coating

Oil pumping was turned on at time t = 5 min at temperature of 27 °C and rate of 0.3 l/min in two days downtime of the bench. At these conditions the soot particles remained on the optical surface (see Fig. 4.32). When output became stabilized the electrical field was applied according to the scheme shown in Fig. 4.35a. As a result of the field effect the soot particle desorption increases and output signal rises (Fig. 4.36, curve 1). The potential difference between the electrodes was about 2000 V and voltage was fed during 5 min with interval of 5 min. The field application leads to accumulated deposit removing from the surface and remarkable increasing output.

Disadvantage of the electrical field cleaning is a necessity to apply high voltage.

Increase of the cleaning efficiency and decrease of the applied voltage value were achieved by using the light guide optical surface as an electrode. It is realized by the use of electrically conductive and transparent coating (with resistance of 15 kΩ and refractive index $n = 2$) on surface of sensitive light guide. ITO coating—composition of indium In_2O_3 (95%) and tin SnO_2 (5%) oxides—was used as the conductive coating. Deposition process was carried out by the evaporation in oxygen medium at 60 °C. In the course of evaluation of the method efficiency the coating was charged positively and the electrodes were charged negatively (Fig. 4.35b). Potential difference was 60 V. The test was performed also in the case of optical rod without coating. Application of the field causes the sharp increase of output up to value corresponding to the absence of the deposit on rod surface (Fig. 4.36, curve 2).

Thus, application of electrical field and use of conductive transparent coating on light guide surface as an electrode allows us to clean the sensitive surface of the light guide while removing the accumulated soot deposit.

The solving of cleaning problem of the optical sensors provides their wide application for condition monitoring of lubricating oils. Particularly, fiber–optical detectors can be built in the oil circulation systems of diesel engines and measure soot content in oil.

4.6 Conclusions

The service life of the engine oil is restricted by thermal decomposition, duration of additive effect, and contamination with soot particles formed due to incomplete fuel combustion. Long operation with degraded oil can reduce considerably the life of an engine. Therefore, up-to-date requirements to diesel engine operation and their reliability set the problem of measuring both soot concentration and its dispersity.

Up-to-date requirements set the necessity of development of portable monitoring devices built in the car lubrication system which allow one to obtain the on-line data on the oil contamination by soot.

To realize on-line soot content measurement for vehicles it is required to produce a measuring system which is relatively cheap to be installed to large number of cars and fairly reliable to operate in a diesel engine. Soot contained in oil varies its electrical (conductivity and permittivity) and optical properties (optical density). Therefore, available devices built in the car lubrication system are based on electrical and optical methods.

Advantages and shortcomings of different devices based on electrical and optical methods are considered and the prospects of practical use of them are discussed above.

References

1. S. George, S. Balla, V. Gautam, M. Gautam, Effect of soot on lubricant oil viscosity. Tribol. Int. **40**(5), 809–818 (2007)
2. R.E. Kornbrekke, P. Patrzyk-Semanik, T. Kirchner-Jean, M.G. Raguz, E.A. Bardasz, Understanding soot mediated oil thickening, Part 6: Base oil effects. SAE Technical Paper 92665, 1998
3. A. Agoston, C. Otsch, B. Jakoby, Viscosity sensors for engine oil condition monitoring—application and interpretation of results. Sens. Actuators A **121**, 327–332 (2005)
4. C. Liu, S. Nemoto, S. Ogano, Effect of soot properties in diesel engine oils on the frictional characteristics. Tribol. Trans. **46**, 12–18 (2003)
5. M. Gautam, K. Chitoor, M. Durbha, J.C. Summers, Effect of diesel soot contaminated oil on engine wear—investigation of novel oil formulations. Tribol. Int. **32**, 687–699 (1999)
6. M. Kaneta, T. Irie, H. Nishikawa, K. Matsuda, Effects of soot on wear in elastohydrodynamic lubrication contacts. Proc. Inst. Mech. Eng. [J] –Part J: J. Eng. Tribology **220** (3), 307–317, (2006)
7. J. Fitch, Four lethal diesel engine oil contaminants. Mach. Lubr. **5** (2007) Available: http://www.machinerylubrication.com/Read/1033/diesel-engine-oil-contaminants
8. S. Aldajah, O.O. Ajayi, G.R. Fenske, I.L. Goldblatt, Effect of exhaust gas recirculation (EGR) contamination of diesel engine oil on wear. Wear **263**(1–6), 93–98 (2007)
9. S. Ashley, Diesel cars come clean. Mechanical Engineering **119**(8), 52–57 (1997)
10. D.D. Troyer, Get ready for more soot. Practicing Oil Analysis **7** (1999) http://www.machinerylubrication.com/Read/51/soot-oil-engine
11. A.D. Clague et al., A comparison of diesel engine soot with carbon black. Carbon **37**, 1553–1565 (1999)

12. R.E. Rosensweig, Ferrohydrodynamics (Courier Corporation, 1997), p. 344
13. L.V. Markova, N.K. Myshkin, H. Kong, M.S. Semenyuk, E.-S. Yoon, H.-G. Han, Up–to–date requirements to efficiency of monitoring of diesel combustion engine oil. J. Friction Wear **23** (4), 60–70 (2002)
14. D.E.G. Crutcher, R. Gervais, L.A. Toms, Use of FT—IR Spectrometry as a replacement for physical property testing of railway lubricant. in Proceedings of a Joint Conf. "Integrated Monitoring, Diagnostics and Failure Prevention", (Mobile, Alabama, 1996), pp. 207—216
15. T.M. Hunt, *Handbook of Wear Debris Analysis and Particle Detection in Liquids* (Elsevier Science, London and New York, 1993), p. 488
16. Soot in diesel engine lubricating oils portable, rugged infrared analyzer. Spectro Scientific Co. Available: http://www.spectrosci.com/product/infracal-soot-meter/
17. J.P. Guillermie, G. Gil, US Patent 3876935, G01R 27/02. Device for Determining the Presence of Impurities in Oil Contained in the Crank Cases of Engines or Other Apparatus—No.05/457,332. Filed 3 Apr 1974. Published 8 Apr 1975
18. C.A. Megerle, US Patent 5089780, G01N 027/413. Oil Quality Monitor Sensor and System—No. 07/663,770. Filed 4 Mar 1991. Published 18 Feb 1992
19. R.A. Bauer, R.W. Hirthe, M.H. Polczynski, M.A. Seitz, J.E. Hansen, US Patent 6278281, G01N 027/02, G01R 027/26. Fluid Condition Monitor—No. 09/220,556. Filed 23 Dec 1998 Published 21 Aug 2001
20. J.P. Heremans, S.-C. Wang, T. Schroeder, L. Nagy, U.S. Patent 6,911,830, G01N 27/06; G01N 33/28; G01N 33/26; G01R 027/08. Diesel engine lubricating oil contaminant sensors method—No. 2004/0036487. Filed 26 Feb 2004. Published 28 June 2005
21. S. Yasuhara, H. Kobayashi, T. Kita, H. Saito, US Patent 4646070, G08B 021/00. Oil Deterioration Detector Method and Apparatus—No. 06/441,035. Filed 12 Nov 1982/ Published 24 Feb 1987
22. A.D. Dickert, E.L. Johnson, J.F. Kirkpatrick, K.A. Hawn, US Patent 5262732, G01R 027/26; G01N 027/74. Oil Monitor with Magnetic Field—No. 07/807,041. Filed 11 Dec 1991/ Published 16 Nov 1993
23. C.E. Freese V, M.J. Rostoskey, R.E. Garvey III, US Patent 5604441, G01R 027/26. In-Situ Oil Analyzer and Methods of Using Same, Particularly for Continuous On-Board Analysis of Diesel Engine Lubrication Systems—No. 08/403,638. Filed 14 Mar 1995. Published 18 Feb 1997
24. C.J. Collister, Patent WO 9850790, G01N 33/28; G01N 27/22; G01R 27/26. Electrical Measurement of Oil Quality—No. WO1998GB01321. Filed 5 July 1998. Published 11 Dec 1998
25. L.L. Nagy, Jr. M.E. Myers, US Patent 4345202G01R 027/04. Method of Detecting Soot in Engine Oil Using Microwaves—No. 06/218,163. Filed 19 Dec 1980. Published 17 Aug 1982
26. S. Kamiya, M. Nozawa, T. Ishida, H. Sasaya, US Patent 4699509, G01N 033/28. Device for Measuring Contamination of Lubricant—No. 06/724,962. Filed 19 Apr 1985 Published 13 Oct 1987
27. M. Nozawa, Y. Toda, K. Naito, Y. Nomura, R. Osaki, US Patent 5548393, G01N 021/15; G01N 033/30. Oil Deterioration Detection Apparatus and Apparatus for Detecting Particles in Liquid—No. 08/269,587. Filed 1 July 1994. Published 20 Aug 1996
28. K.M. Park, N. Nguyen, M.A. Nassar, US Patent 5907278, B60Q 001/00. Capacitive Sensor Assembly with Soot Build-Up Inhibiting Coating—N0. 09/061,635. Filed 16 Apr 1998. Published 25 May 1999
29. D.P. Ames, D.L. Rode, US Patent 6043505, G01N 015/06. Device and Method for Monitoring Fluids with a Detection of Cross Sectional Shape of Transmitted Beam—No. 09/130,330. Filed 6 Aug 1998. Published 28 Mar 2000
30. V.I. Busurin, Yu.R. Nosov, *Optic-fiber detectors: Physical basis, problems of development and application*, (Energoatonizdat, Moscow,1990), p. 256 (in Russian)

31. H. Kong, H-G Han, E-S Yoon, L. V. Markova, M. S. Semenyuk, U.S. Pat. No. 6842234, G01N 033/28. Apparatus for measuring soot content in diesel engine oil in real time—No. 10/291,113. Filed 8 Nov 2002. Published 11 Jan 2005
32. A.D. Zimon. *Adhesion of Dust and Powder*, (Plenum Press, 1969), p. 424
33. C.V. Ossia, H. Kong, L.V. Markova, N.K. Myshkin, Optical absorption under total internal reflection in the characterization of soot in diesel engine oils. Meas. Sci. Technol. **20**(1), 015205 (2009)

Chapter 5
Wear Prediction for Tribosystems Based on Debris Analysis

5.1 Introduction

Most of machines and mechanisms operating with oil can be successfully monitored by wear debris analysis. Wear debris parameters relate to the mode and severity of wear of the rubbing solids. While the size and concentration of debris are reliable indicators of the machine state, their morphology and composition can give information on the wear mode and the source of debris generation. Moreover, the concentration of particles in the oil sample and their size distribution characterize the wear rate and the degree of wear [1, 2].

Wear is the process of mass loss (Δm) evolving in time. The wear of a friction pair is presented graphically as a function of time (or friction path) at constant load and sliding velocity conditions, the slope of the curve in each point characterizes the wear rate. Three stages of wear usually exist which are shown in Fig. 5.1. The first portion of the curve (first stage I) is the initial period of wear which lasts shortly compared with the expected life of a friction unit. This stage is characterized by the unsteady state of the tribosystem with a high wear rate, which is gradually reduced. It is termed the running-in period when rubbing surfaces adapt to each other, particularly while the so-called equilibrium roughness is reached. Friction and wear modify in such a manner that the tribosystem transforms into the steady state.

Second stage II follows and it lasts longer being characterized by stable conditions of friction and practically constant and relatively small wear rate. This stage occurs as a normal operation of friction unit, wear grows almost linearly causing minor damage of the surface. Accumulation of wear alters the shapes of rubbing pairs and the clearance between them, leading finally to the period when the wear rate increases strongly and severe wear begins (stage III).

N.K. Myshkin and L.V. Markova, *On-line Condition Monitoring in Industrial Lubrication and Tribology*, Applied Condition Monitoring 8,
DOI 10.1007/978-3-319-61134-1_5

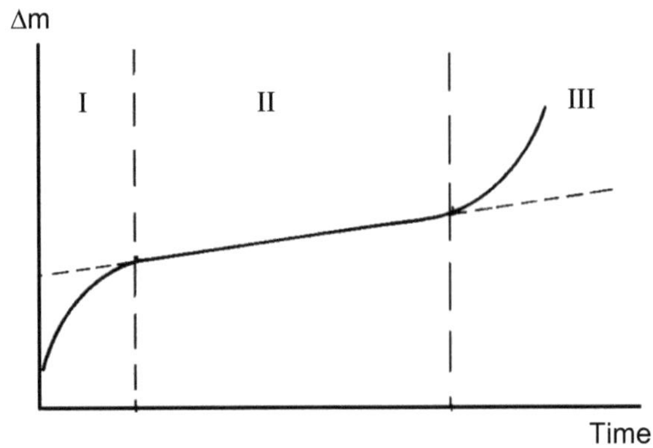

Fig. 5.1 Typical stages of wear process

Table 5.1 Correlation between wear stage and parameters of wear debris

Wear debris classification based on references	Wear stage		
	Running-in	Steady-state (normal)	Catastrophic (severe)
[100, 101]	Spheres with the diameter of 1–20 μm, Platelets and flakes	Platelets and flakes with length <25 μm, Smooth platelets with the length to 15 μm and thickness to 0.25–1.0 μm	Chunks, Long platelets with surface scores and length >20 μm, Chips with the length of 5–100 μm and thickness of 2–15 μm
[102]	–	Particles with size <15 μm, average size of 2 μm	Particles with size <150 μm, average size of 20 μm

It should be noted that real tribosystems operate with stops, during re-starts they run-in again, yet to a smaller extent.

Wear debris are formed in operation of the tribosystem being the result of damage in the friction contact. Table 5.1 presents the parameters of debris relating to running-in, steady-state wear, and severe wear stages. It is evident that particles of wear have shapes and sizes characteristic to each wear stage.

In transition from steady-state wear to severe stage the concentration of wear particles grows and their size distribution shifts to the large sizes (Fig. 5.2), [3]. Henceforth, the sudden appearance of large particles in the oil indicates the onset of severe wear.

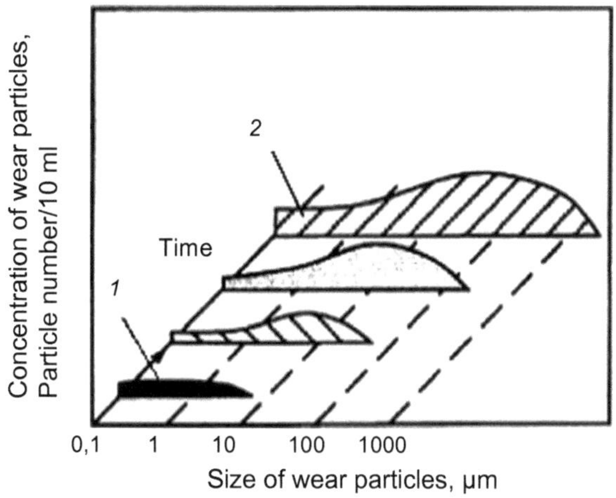

Fig. 5.2 Change in concentration and size distribution of wear particles in the course of tribosystem operation: *1* steady-state wear, *2* catastrophic wear [3]

Wear of tribosystem is considered as a combination of processes on mating surfaces resulting macroscopic behavior of a friction unit. Different mechanisms (modes) of wear can be realized simultaneously and there are different classifications of these modes, but most common can be listed as following: abrasive wear, adhesive wear, fatigue, and oxidative wear.

Abrasive wear results from plowing by hard asperities and abrasive particles forced against a solid surface. It is defined as damage to solid surface due to progressive loss of material in relative motion of the surface and can be severe in most cases.

Adhesive wear results from shear of adhesion bonds generated in friction. Specific character of this mechanism of wear is the material transfer between contact surfaces leading in severe case to seizure. But in many cases material transfer of the softer component can result in normal mild wear with low rate.

Fatigue wear results from cyclic stresses lower than strength limit in friction with further cracking and debris generation.

Oxidative (corrosive) wear results from removal of the oxides films which are formed by interaction of the deformed metal layer with the environment and lubricants.

For any particular tribosystem under certain conditions of friction we can find the dominant mechanism of wear which determines its intensity, but also find one or more additional mechanisms. For example, at dominant fatigue wear the abrasion and oxidation modes can be observed. Hence, a few processes of different origin can be under way simultaneously generating the wear particles.

Usually the wear mechanism and component generating the wear particles are identified by laboratory techniques (ferrography, filtration, and microscopy). They are based on studies in morphology of particles and their composition while using the atlases classifying the particles by texture, shape, color, and size with relation to origin. There is an internationally accepted "Wear Particle Atlas" presenting such information [4].

According to the atlas particles of flat platelets with the length to 15 μm and thickness to 0.25–1.0 μm can be classified as debris of mild adhesive wear. The typical length to thickness ratio for particles with size of 0.5 μm is equal to 3:1 and they are formed at delamination of metal layer at normal wear. This conclusion of the atlas is supported in literature on ferrography where the plate particles with size of ≤ 5 μm were most common in adhesive wear.

Abrasive wear is characterized by cutting-like particles and long strips with the length up to 25–100 μm, width of 2–5 μm, and thickness of 0.2–0.25 μm.

Fatigue wear generates spherical particles with the size >3 μm which are common on ferrograms. Mechanism of their generation is related to separation of large particles from fatigue cracks and further local temperature flash and melting to spherical shape [5–8]. Large particle can also be rolled in the contact with forming of strip-like plates and spheres as well as welding of small particles into large ones casing the abrasion of the surfaces in contact [9].

Let us consider the example of correlation of the wear mechanism and parameters of wear particles in a rolling bearing [10]. Surface damage in such a bearing starts with generation of microcracks in the subsurface layer being formed near the material defects or foreign inclusions in the bearing steel. At stress growth the subsurface crack propagates separating the material and forming the fatigue wear particles. If these particles are hard and comparable with the clearance in the bearing they can increase the probability of the bearing failure. There are four shapes of wear particles which are typical in rolling bearings: microspalling ones, laminar particles, chunks, and spheres. Figure 5.3 shows these particles on filtergrams.

At the early stage as a result of *microspalling* process the particles are formed and rolled further to plate-like particles (platelets) with a smooth surface, length from 10 to 50 μm, and thickness of the ten times less (Fig. 5.3a). Original microspalling particles are often found with the spherical ones being the products of the early stage of fatigue.

Most common of the fatigue wear particles are the *laminar* ones (Fig. 5.3b). Large wear particles can be crumbled into the smaller platelets, and then being rolled in the contact. The laminar particles have the evident peculiarities: they are metal particles with the length from 50 to hundreds of micrometers, sometimes even to few millimeters; the ratio of diameter to thickness is in the range from 20:1 to 50:1.

Further result of surface pitting and spalling is presented by large *chunks* of fatigue wear. At this stage fatigue cracks propagate inside the subsurface layer and it is a real threat of the bearing failure. The appearance of large chunky particles can be used as an indicator of fatigue failure of the rolling bearing. Such particles are

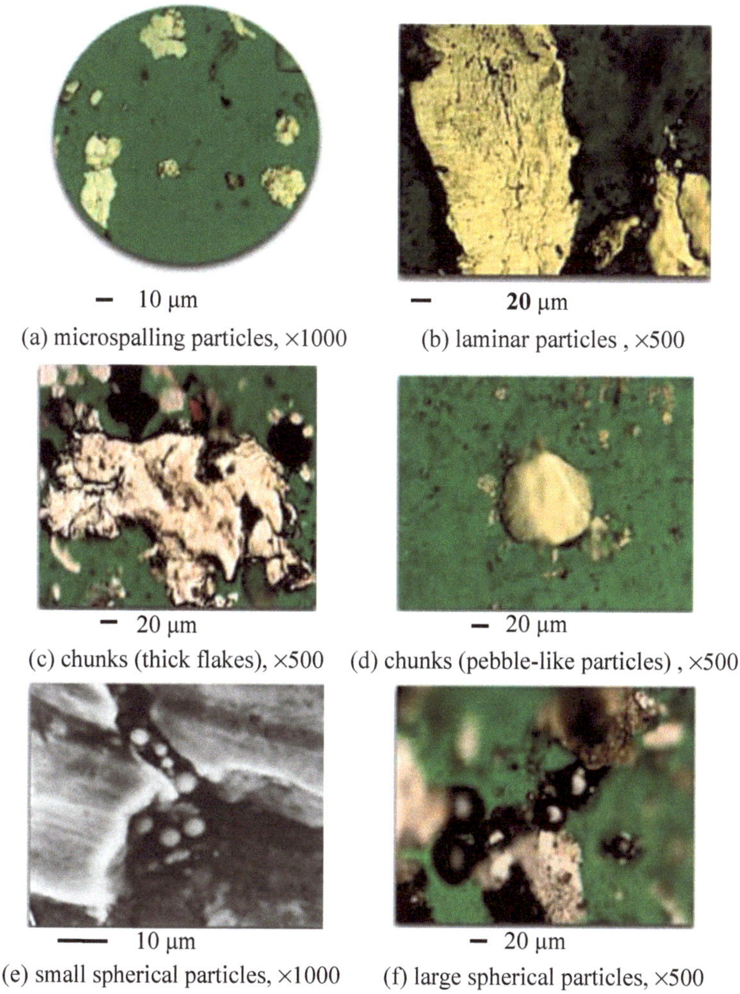

Fig. 5.3 Typical wear particles on filtergrams [10]

formed at deep scaling being thick platelets produced by multiple rolling. Their thickness is from 5 to 20 µm or more and the ratio of sides equals 10:1 as seen in Fig. 5.3c. In case of deep scaling the contact surface is severely damaged. At this stage thick pebble-like particles are found which have not been rolled many times. Such particles are easily recognized visually due to their shape being in all dimensions between dozens and hundreds micrometers with the ratio from 5:1 to 1:1. Figure 5.3d presents the pebble-like particles found in the lubricated rolling bearings.

The most valuable indicator of the fatigue wear severity is the occurrence of *spherical* particles, their size and quantity. Such particles are generated in different mechanisms of wear and at various stages, but usually the small spherical particles (less than 5 μm) are related to rolling bearing fatigue while larger ones can be related to sliding, plowing or cavitation. Small spheres are mainly formed in fatigue cracks (Fig. 5.3e) and they indicate the beginning of surface pitting. Large spherical particles (more than 10 μm) results from seizure, welding, plowing etc. If they are mixed with large flakes and/or chunks that means the occurrence of deep crumbling of the surface layer. In that case large particles often have the signs of overheating and melting (Fig. 5.3f).

Because the large spherical particles are produced at concentrated metal-metal contact at high temperature their presence at rolling can be an additional indicator of wear severity. For example, wear at slip at rolling followed by generation of large spherical particles is more severe than wear without forming such particles indicating high contact temperature.

Above-mentioned research data clearly illustrate the correlation of wear particles parameters and processes occurred in a friction contact.

Attempts of different laboratories to connect the wear debris parameters with the mechanism of wear in tribosystems result in classifications based on various concepts.

Several works [1, 11–16] have been devoted to the question of the classification of wear particles on the basis of their morphology (size and shape) in accordance with the basic wear mechanisms, and these different classifications are determined by the specific test pair type and the process conditions. Thus, as the result of tests performed on a friction machine the authors of [12] identify the following six wear modes based on wear particle size: hydrodynamic (wear particle size about 5 μm), boundary friction (<15 μm), friction with breakdown of the lubricant film with traces of scuffing (<150 μm), oxidative wear (<150 μm), and catastrophic wear (<1000 μm). However in [13] in testing a gear train the following wear regime classification is established, based on the wear particle size: normal wear (particle size up to 15 μm, maximal number of particles of size about 2 μm) and catastrophic wear (particle size up to 150 μm, the majority of the particles are of size 15–25 μm).

There are various wear particle classifications based on shape. In tests on a four-ball machine six particle types are identified in [11] ("grains," "platelets," "flakes," "chunks," "spheres," and "laminar particles"), corresponding to the different wear regimes. In several cases [12, 17, 18] analysis of the chemical composition of the wear particles is used in studying the wear mechanism. Materials containing iron—low-alloy and high-alloy steels and gray irons—are widely used in the fabrication of the parts of friction assemblies. Mechanical contact of these parts in the severe friction regimes is accompanied by significant temperature increase in the contact zone, which leads to the appearance of iron oxides. Tests of low-carbon chromium steels showed that in the oxidative wear regime the wear particles consist primarily of α–Fe_2O_3. In the more severe friction regime the wear products are a mixture of the iron oxides (γ–Fe_2O_3, Fe_3O_4, FeO) [12].

The results of study of the magnetic properties of the wear particles were presented in [18]. It was found that in the lubricants of iron-containing friction pairs there may be both ferromagnetic iron components α–Fe, Fe_3O_4, γ–Fe_2O_3, Fe_3C and paramagnetic iron components $FeSO_4$, $FeSO_4$-$7H_2O$, $FeCl_2$, FeO, $FeCl_2$-$4H_2O$, which must be considered when using the magnetic methods of wear particle analysis.

The Standard ASTM D7684–11 [19] proposes two approaches for classification of wear debris extracted from in-service lubricant and hydraulic oil samples. The first approach plans the finding of a group number to which test wear debris are related basing on their main features according to the Table 5.2. In the table "aspect" means the ratio of orthogonal minimum to maximum dimensions for the two-dimensional image of particle. Then, the wear severity is defined using this group number.

The second particle classification approach is based only on particle size and shape. Particle type is identified using the Table 5.3 data.

Due to the fast progress in computers and methods of image recognition it has become possible to replace the visual microscopic analysis by automatic one [20–23]. Three instruments including laser scanning confocal microscopy (LSCM), scanning electron microscope (SEM) and atomic force microscope (AFM) have been applied to investigate 3-D particle information.

The automatic system was developed which can classify 6 types of metallic wear particles: normal wear debris, cuttings, spherical, flake-like, chip-like, and severe slip particles [20]. Laser scanning microscopy has been used to get 3D images with further digital processing. Morphology (size and boundary shape) and surface topology (roughness and texture) were transformed into nine parameters where five of them were related to morphology, three ones to topology, and one parameter to value of each parameter in the system of features for particles of different types. Such a system can replace the expert estimates, shorten the time of monitoring tribosystem, and prevent the errors in expert visual estimates. Methods of 3D surface image analysis of wear particles and its digital interpretation by fractals are considered elsewhere [21].

Since the particle is fixed on the slide, these instruments can only capture the particle information on one side, which means the particle thickness is still partially obtained.

Recently video based methods were developed to investigate more detailed particle features [24, 25]. The wear particles pass through a flow channel, where the camera is mounted to capture the video of particle in the flowing lubricant. Features of individual particle such as colour has been extracted to investigate the particle oxidation. In addition, this methodology has also been applied to estimate 3-D particle information. However these methods can not be used for on-line wear particle identification in particular because they need manual operation.

To create reliable methods of on-line automatic wear particle recognition new methods are under development. A prototype of on-line extraction and

Table 5.2 Classification of wear particles by their main features

Group number	Concentration	Size	Aspect	Shape	Color	Texture	Composition	Classification	Severity
1	Few	Fine, <6 μm	1:1	Platelets	Red	Bright or reflective	Ferrous metal	Abrasive wear	Normal
2	Moderate	Small, 6–14 μm	1:2	Ribbons	Black	Dull or oxidized	Cupric metal	Mild sliding wear	Low alert
3	Many	Medium, 14–40 μm	1:3	Chunks	Tempered	Pitted	Other metal	Severe sliding wear (metal removal)	High alert
4	Dense	Large, 40–100 μm	1:10	Spheres	Metallic	Striated	Dust	Rolling fatigue wear (subsurface spall)	Low fault
5		Huge, >100 μm	1:100	Spiral	Straw	Smeared	Organic	Corrosive wear	High fault

Table 5.3 Decision table for classification of particles

Particle Type	Size, Major Dimension	Shape Factor (Major/Minor Dimension)
Rubbing wear particles	<15 μm	Thin, >5:1 Usually about 10:1
Rubbing wear particles	<5 μm	Any shape except curved or curled
Abrasive wear particles	Any size	Long, thin, curled or curved, ribbon-like
Severe wear particles Chunks	>15 μm >5 μm	>5:1–<30:1 <5:1
Reworked (laminar) particles	>15 μm	>30:1

three-dimensional characterisation of wear particle features from video sequence is newly proposed by Wu et al. [24]. The proposed method is based on analysis of particle multiple views and includes the following main procedures: (1) dynamical particle extraction with Gaussian mixture modeling; (2) multiple particle tracking for information extraction from individual particle data; (3) volumetric reconstruction of wear particle from multiple views. Compared with other particle imaging approaches, this method can provide the thickness information as well as particle volume which can be used to estimate the material loss and wear rate. The automatic process allows fast 3-D particle characterization, and make it a potential option for on-line monitoring.

Information in wear particle atlas is rather abstract, not fixed to a particular mechanism and machine. In practice, preliminary information on wear debris size and concentration at normal, severe, and catastrophic wear in a particular mechanism is very important in condition monitoring, especially, in real time monitoring using built-in devices in order to provide reliability of wear sensor and alarm signal. This information should be obtained on laboratory equipment and using it in test operation a suitable method of condition monitoring which will record deviation from normal wear mode can be chosen. The more full information on correlation of the wear mode and parameters of debris is used at development of wear monitor sensor, the better monitoring result will be.

However, in practice inventors of wear monitor sensors do not have sufficient preliminary information. Most industrial companies have neither time nor money to initiate a complete survey of monitoring devices that might suit them best. Therefore, engineers are to use indirect information on friction testing on laboratory bench under load conditions similar to real ones. Some rough data can be taken from technical publications on a similar machine operation. The information on some case studies of particular mechanisms is presented in Table 5.4 being composed of published data.

Table 5.4 Parameters of wear debris generated in machines

Machine type	Chemical composition of wear particles	Wear particle size, μm		Wear particle concentration, wt.%		Reference
		Normal operation	Catastrophic wear	Normal operation	Catastrophic wear	
1	2	3	4	5	6	7
Aircraft						
Engine	Ferrous	10–15		$(1–6) \times 10^{-4}$	$(12–16) \times 10^{-4}$	[1]
Transmission	Ferrous			$(2–6) \times 10^{-4}$	$(18–105) \times 10^{-4}$	[2]
Combat aircraft						
Engine F119-PW-100	Ferrous	>100				
Hydraulic oil	Non-ferrous	15–100				[100]
Engine F18, F404 with three micron oil filter	Al, Cu, Pb, Zn					[101]
Helicopter						
Gearbox	Fe Zn, Fe, Mg, Si, Cr, Al, Cu, Ag, Pb, Ni	5–10		Zn: $(6,6–7,4) \times 10^{-4}$, Fe: $(4,1–7,8) \times 10^{-4}$, Mg: $(3,8–12,7) \times 10^{-4}$, Si: $(1,2–3,8) \times 10^{-4}$, Cr: $(0,3–0,6) \times 10^{-4}$, Al: $(1,4–3,1)) \times 10^{-4}$, Cu: $(1,9–3,6) \times 10^{-4}$, Ag: $(1,9–2,0) \times 10^{-4}$, Pb: $(0,8–1,2) \times 10^{-4}$, Ni: $(0,6–1,5) \times 10^{-4}$,		[4]

(continued)

Table 5.4 (continued)

Machine type	Chemical composition of wear particles	Wear particle size, μm		Wear particle concentration, wt.%		Reference
		Normal operation	Catastrophic wear	Normal operation	Catastrophic wear	
1	2	3	4	5	6	7
Automobile						
Engine	Ferrous Fe, Al, Cr, Cu, Pb	<15	>15 >100	Fe: 28×10^{-4}, Al: 7×10^{-4}, Cr: 1×10^{-4}, Cu: 14×10^{-4}, Pb: 21×10^{-4}	Fe: 160×10^{-4}, Al: 29×10^{-4}, Cr: 11×10^{-4}, Cu: 154×10^{-4}, Pb: 408×10^{-4}	[5] [6]
Bearings	Al, Zn, Pb			Al: $(5–15) \times 10^{-4}$		[5]
Piston rings (Cr-coated surfaces)	Fe, Cr			Fe: $(2–9) \times 10^{-4}$, Cr: $(2–18) \times 10^{-4}$,		
High Speed Train						
Diesel engine	Al Na			$<10 \times 10^{-4}$ $<90 \times 10^{-4}$	20×10^{-4} 300×10^{-4}	[7]
Steam turbine-generator						
	Fe, Si, Ca, Cu, Al, Ba, K, S, Mg, Cl, P	1–10	1–20			[8]

5.2 Laboratory Diagnostics Basing on Wear Debris Analysis

The history of the development of wear diagnostics based on the wear products goes back only four decades. However in this relatively short period the path has been traveled from qualitative description of the wear particles in the various regimes to quantitative estimates of the characteristics of the wear process.

Methods for diagnosing the machine state based on examining wear debris involve two groups: methods for off-line (external) examination and those using devices built into oil circulation system [26] (Fig. 5.4). The methods of the first group are usually used in laboratory analysis and involve periodic sampling of the oil and its further examination. These methods are realized in rather complex laboratory apparatuses or portable devices employed in field tests. The built-in devices permit to monitor the machine condition in real time.

Classification of various monitoring methods and comparison of their capabilities are shown in the Table 5.5 [27]. *We* shall examine in sequence these groups of methods.

In this group the methods most widely used for wear diagnostics are the optical methods, among which the methods of atomic and molecular spectroscopy occupy a special place.

Emission spectrophotometry. This method has been used since the 1950s as a method for diagnosing the condition of internal combustion engines [28]. The intensity of wear of tribosystems is indirectly characterized by the concentration in

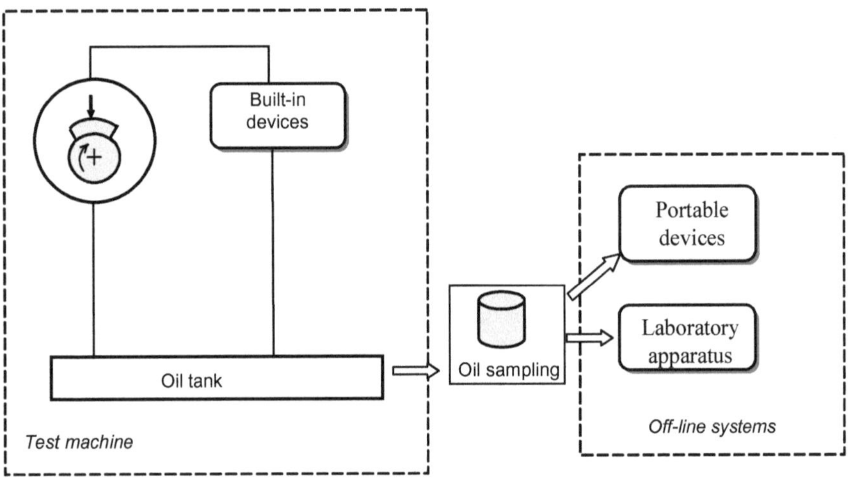

Fig. 5.4 Condition monitoring systems based on wear debris analysis

Table 5.5 Laboratory methods of monitoring the lubricated tribosystems [27]

Analysis method	Characteristics					Continuity of monitoring	Use of express-devices
	Composition of particles	Particle concentration in medium	Fractional	Morphology	Total oil contamination		
Emission spectrophotometry	+	+	–	–	–	–	–
Atomic-absorption spectrometry	+	+[a]	–	–	–	–	–
Atomic-fluorescence spectrophotometry	+	+	–	–	–	–	+
Absorption spectrophotometry	–	+	–	–	+	–	+
Microscopy	–	+	+	+	–	–	–
Ferrography	+[b]	+	+	–	–	+	+
Chromatography	–	+	+	–	–	–	–

[a]For particles smaller than 10 μm
[b]Use of bichromatic microscope make it possible to determine the presence of oxides

the oil of metals, which are present in the materials of the rubbing parts. Emission spectral analysis is based on the spectra of atoms and ions excited by electromagnetic radiation. The basis of the emission spectrophotometric systems is the "internal standard" method, involving measuring the ratio of the intensity of the line of the element being analyzed to the comparison line. The beam of light from the arc or spark is limited by a slit and is broken down into spectrum with the aid of diffraction lattice or quartz prism. The light beams corresponding to the wavelength emitted by the elements being analyzed are separated from the spectrum by the exit slits and are projected with the aid of mirrors onto photo receivers for further comparison. The application of emission spectrophotometry for wear monitoring has been covered quite completely in [29, 30]. With the objective of improving the analysis accuracy, reference [31] proposes the plotting for each wear product of calibration curves independent on the presence of additives, viscosity or type of the base oil. In this case, the accuracy increases by five times in comparison with the conventional spectrometry.

Emission spectral analysis is performed both for liquid oil specimens and for oil specimens with preliminary ashing of the sample. In the later case the sediment is first separated from the oil sample and the ash residue is burned in spectrometer. This method has high sensitivity in determining small concentrations of elements in used and unused oils. The development of multichannel emission spectrophotometric systems makes it possible to perform large-scale mutlicomponent analyses of the used oils. This method is most widely used in engine diagnosis under laboratory conditions. However, the process of preparation of the oil sample for analysis makes it difficult to carry out continuous monitoring of the engine condition.

Nowadays most emission spectrometers employed in used oil analysis are based on either of the rotating disk electrode (RDE) technique or of the inductively coupled plasma (ICP) technique [32, 33].

In the RDE technique, a rotating disc electrode brings an oil sample into the gap between the disc and stationary rod electrode. A high voltage arc between the disc and rod electrodes causes the individual atoms in the sample to give off light or radiant energy.

The ICP excitation technique employs electrode-free plasma created by a flow of inert gas (argon). The ionized argon reaches temperatures of 8000–100,000 °C. The oil sample is aspirated into the plasma where the atoms are totally dissociated and excited sufficiently to give off radiant energy. In RDE and ICP systems, a lens or fiber optic is used to gather and focus the radiant energy onto a concave diffraction grating which disperses the light into various spectral lines associated with the elements. Photomultiplier tubes are positioned behind the exit slits to detect and convert radiant energy into electrical signals to measure the concentration of each metal present in the sample.

For lubricating oil analysis RDE technique is preferred for its simplicity and reliability and RDE spectrometers are responsive to relatively large particles, but the upper limit has traditionally been only around 10 μm.

Atomic absorption spectrometry (AAS). In this spectral analysis method the sample is vaporized in an atomizer (in a flame or plasma), then we measure the intensity of the light from a discrete radiation source, passing through the vapor of the sample being studied. On the basis of the degree of attenuation of the intensity of the lines of the element being determined we evaluate its concentration in the sample [33, 34]. A technique for determining the iron and copper content in oils by the atomic absorption method is following. The sample is calcined in an electrical muffle furnace and then treated with acids. The solution of the ashy sediment in hydrochloric acid is subjected to analysis on a spectrophotometer. The content of the metals in the analyzed solution Is found from calibration charts. This technique is used to evaluate the anti-wear properties of motor oils, and also in determining the technical condition of engines. Analysis of the effectiveness of determination of the wear product content in the oil by the AAS method showed that this method reacts primarily to the small particle; suspended in the oil sample, and does not react to large particles of dimension more than 10 μm.

Atomic fluorescence spectrophotometry. In this method the vapors of the sample being studied are irradiation by radiation that is resonant for the sought element and Its fluorescence is recorded. Roentgen radiation is normally used to irradiate the oil sample [35–37]. With the use of this method there are no limitations on the particle size, and we can rapidly determine the presence in the lubricant of large wear particles characterizing the onset of catastrophic wear. The authors of [35] used this method to diagnose the condition of machines on the basis of the iron content in the oil. Atomic fluorescence spectrophotometry is applicable only for those metals which have realizable resonance radiation. The X-ray fluorescence method can serve for determination of the same elements as the AAS method, except for lithium and magnesium [36].

It should be noted that absorption of light by additives in oil brings the error in results, so in order to improve reliability, the analysis of particles deposited in 0.8 μm filter is employed [38].

Absorption spectrophotometry. This method is based on the Bouger-Lambert-Beer law and establishes the connection between the intensities of the incident light and the light passing through the substance as a function of the absorbing layer thickness and the substance concentration. It makes it possible to determine the wear particle concentration in used oil [39]. Measurements of the optical density are normally used for the analysis of used oils, see, for example the express-method for determining the impurities in used oils [31]. While making the analysis, solution of the used oil and reference one of unused oil are prepared for increasing the transparency of the sample. From the comparison of the optical densities of these solutions the concentration of insoluble impurities in the oil can be found. This method makes it possible to evaluate the effectiveness of oil filters and the quality of the engine operation. However, the need for additional sample preparation and inadequate accuracy limit its wide application.

In addition to the spectral analysis methods, microscopy is widely used in tribodiagnostics.

Optical and electron microscopy. This method makes it possible to count the number of wear particles and also determine their size distribution [12, 40]. The wear mechanism is classified in accordance with the number, dimensions, and shape of particles which are studied in optical and electron microscopes [12]. The oil sampling methods and the technique for counting the number of particles and determining their dimensions were described in details [40].

In addition to microscopy *the magnetic methods* are widely used in tribosystem condition diagnosis based on the wear particles. The magnetic methods can be used to determine the paramagnetic and ferromagnetic components of the oil. In used oil there may be paramagnetic iron compounds $FeSO_4$, $FeSO_4 \cdot 7H_20$, $FeCl_2$, FeO, $FeCl_2 \cdot 4H_2O$ and ferromagnetic components α-Fe, Fe_3O_4, γ-Fe_2O_3, Fe_3C [18].

The analytical ferrography method proposed by Seifert and Westcott [14] is widely used for off-line analysis of the oil sample.

Analytical Ferrography makes it possible to study the size distribution of the ferromagnetic and paramagnetic particles of used oil. Ferrography has been used as a wear particle study method since the 70s ana Is described in detail in [1, 2, 40–43]. It finds application as a means for predicting damage and preventing failure of mechanisms because of premature wear in such engineering fields as jet engines, diesel engines, helicopter transmissions, hydraulic systems, and so on. The analytical ferrography method includes two stages: deposition of the particles in a magnetic field (obtaining the ferrograms); information processing (analysis of the ferrograms).

Deposition of the ferromagnetic and paramagnetic particles from the lubricant flow is accomplished under the influence of a high magnetic field gradient, directed vertically downward. The force acting on the particle is proportional to the wear particle volume and its magnetic permeability.

The deposition of particle in the laminar flow is described by the equation:

$$6\pi\eta a\frac{dz}{dt} = \mu\,\mathrm{grad}_z H + (\rho_p - \rho)\frac{4}{3}\pi a^3 g,$$

where η—viscosity of oil; *a*—particle radius; H—magnetic field intensity; μ—particle magnetic moment (when $H > I_s$,); I_s—saturation magnetization of magnetic material; ρ_p, ρ—densities of particle and medium; g—acceleration of gravity.

The deposited particles arrange themselves on the surface of the substrate in chains that are perpendicular to the direction of lubricant flow, with the large particles located at the beginning of the substrate and the small particles located at the end. After the sample drains from the substrate a solvent and a fixing solution are passed, and after drying the ferrogram is ready for further study. The ferrograms are analyzed by measuring their optical density on the glass substrate, by the direct readout method, or with the aid of the image analysis system. The latter can measure dimensions of the wear particles along the ferrogram with microscope coupled with a scanning device that makes possible to feed the data to computer. It is also possible to study the ferrogram directly in a bichromatic microscope and determine the presence of oxides, sulfides, and clean metals, or heat the ferrogram, and again observe the particles color (e.g. at 330 °C the particles of the low-alloy steel alloy become bluish).

Study of nonmagnetic solid particles can be realized by the use of a nonuniform electrostatic field, the gradient of which does not coincide with the magnetic field gradient. In studying the wear particles of nonmetallic materials it is possible to use special fluids, containing oil solvents which are miscible with water and magnetic salts that are adsorbed on the wear particles. The obtained solution is studied by the conventional ferrography method.

Chromatography. The method involves separation of the components in liquid phases; in the first stage of the chromatographic study the multicomponent liquid system is fractionated, and in the next stage the separated components are analyzed. Thus, the dispersing properties of lubricating oils are evaluated with the aid of filter paper chromatography [44]. A drop of the oil being tested is applied to the filter paper sheet alongside a drop of fresh oil, to which there is added a definite amount of lampblack. Soaking into the paper, the oil forms a characteristic circular spot, in which four zones are identified: inner, boundary, dispersion, and oily. The larger the radius of the dispersion zone, the better is the quality of the oil. The chromatogram is analyzed with the aid of a photoelectric photometer and the light signal is converted to an electric signal. The latter is fed to a recording voltmeter. This technique has been developed which makes it possible to determine the quality and degree of contamination of the oil from the shape and characteristic dimensions of the curve.

The data obtained in laboratory study by different methods (spectroscopy, feirography, light scattering, flow ultramicroscopy, activation methods, etc. are fairly comprehensive for adequately estimating the mechanism conditions. Thus, British Rail Company applies spectroscopy when monitoring the diesel engines. Shell Company applies gravitometry and spectroscopy to estimating the machine state. However, laboratory analysis necessitates expensive equipment and skilled operators which results in its high cost. In addition, the time interval between sampling and obtaining the data on machine state is often inadequately long. In this connection there is a trend of producing portable devices for express-analysis which can be used at service and repair centers.

5.3 Portable Means for Machine Condition Monitoring

The main destination of the portable devices for machinery monitoring is to provide information for predictive maintenance in advance on-cite, at service center or monitoring laboratory. These devices must be simple in operation and relatively cheap, while providing reliable data. Among the devices available nowadays we can distinguish the devices based on magnetic measurements,—Direct Reading Ferrograph, Rotary Particle Depositor, Particle Quanifier, and Wear Particle Analyser.

Direct Reading Ferrograph (DRF) was developed on the base of analytical ferrography in order to carry out analysis of ferrograms without complicated laboratory equipment. As a result of modification the DRF has become available for usage not only in research laboratories but in service centers. It consists of magnetic system, light source, electronic block, and indicator (Fig. 5.5) [32]. The sample of oil is pumped through the glass pipe mounted inside the magnetic system and wear particles are deposited on the pipe wall within two zones of strong gradient of magnetic field: zone L for large particles and zone S for small ones. Optical density of deposits is estimated by change in light coming through the zones and then through the light guides to the electronic block.

Electronic block performs a signal processing consisting in calculation of the wear index W, which is usually expressed by the relation [1, 2]:

$$W = (D_L + D_S)(D_L - D_S) = D_L^2 - D_S^2,$$

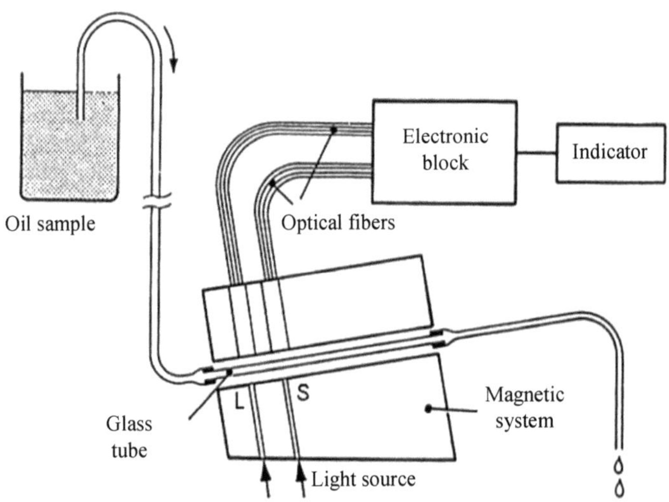

Fig. 5.5 Scheme of direct reading ferrograph [32]

where D_L, D_S are the optical densities of the precipitate at the beginning and end of the ferrogram; $(D_L + D_S)$ is the wear particle concentration index (characterizing the wear rate); the quantity $(D_L - D_S)$ is the index of the degree of shift of the particle size distribution toward the region, of large particles (characterizing the degree of wear). The magnitude of the wear index W depends on the degree of dilution of the sample, which is taken into account with the aid of the dilution factor.

The ferrography method is widely used to determine and predict the wear *of* jet engines, diesels, helicopter transmissions, hydraulic systems, and so on [16, 45, 46].

In spite of its wide use, ferrography is characterized by the following drawbacks. A large part of the wear products deposits on the first 5–10 mm of the substrate, which sometimes leads to significant compaction of the deposit and hinders study of the individual particles. High dilution is required in order to obtain a deposit suitable for study in case of highly contaminated samples. This dilution reduces the validity of the analysis. In studying oils that are highly contaminated with non-ferromagnetic particles the deposit becomes also contaminated, which decreases the reliability of test results.

These drawbacks were overcome in the Rotary Particle Depositor (RPD), developed in Swansea Tribology Center (Great Britain) in the 1980s. The system operating principle is based on the simultaneous action of gravity and magnetic fields on the oil sample, placed on the substrate [47]. RPD consists of magnetic system *1* (Fig. 5.6a) with attached deposition substrate *4* and rubber seal *2*. The magnetic system includes two co-axial magnets—internal cylinder magnet *5* with the upper north pole and outer ring magnet 6 with the upper south pole (Fig. 5.6b). Lines of magnetic field form three zones with maximal gradient. The oil sample is slowly deposited in the center of the substrate by the pipette *3* at velocity of rotation equal 70 rpm. Then the sediment is washed by solvent from the other pipette at rotation velocity equal 130 rpm and dried at 200 rpm during 1–2 min.

The special configuration of the magnetic field and the selection of the rate of rotation of the substrate with the sample lead to deposition of the particles present in the oil in three concentric circles. The large particles (>50 μm) deposit in the inner ring together with some part of the fine particles. Particles of intermediate size (10–50 μm) and a few particles of dimension less than 10 μm are deposited in the second ring. The remaining finer particles are found in the outer ring. A large part of the nonmetallic particles deposits in the outer (third) ring. After deposition of the particles the deposit is flushed with solvent and then dried. Then it can be subjected to qualitative and quantitative analysis, just as in ferrography. The RPD is used efficiently in monitoring the condition of gearboxes, diesel engines, aircraft piston engines, and hydraulic systems.

A specially developed "Particle Quantifier" (PQ) [48], consisting of an alternating current magnetometer, is normally used for quantitative analysis of the precipitate obtained with the aid of the RPD. PQ includes a couple of differentially balanced coils *4, 5* (Fig. 5.7) which are placed in magnetic field of the coil 2 fed by the AC with the frequency 60 kHz from the generator *3*. The coils 4 and 5 are balanced by resistors R1, R2 in order to get zero differential signal of the e.m.f.

Fig. 5.6 Rotary particle depositor: **a** *1* magnetic system, *2* rubber seal, *3* pipette, *4* deposition substrate; **b** lines of magnetic field, *5* internal cylinder magnet, *6* outer ring magnet [47]

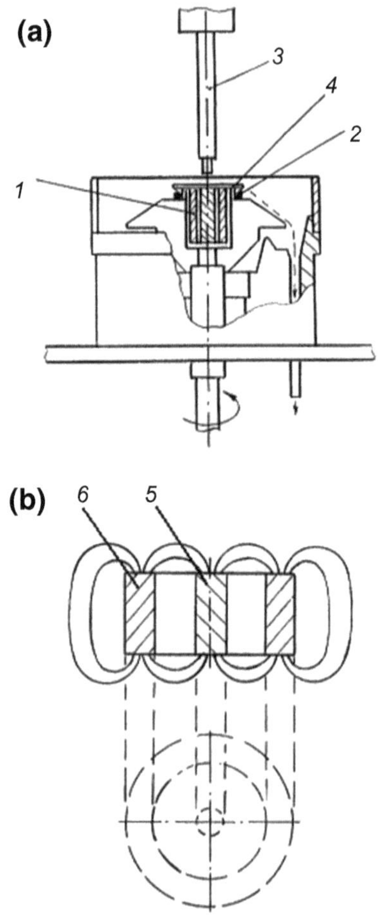

generated in the coils. The balancing is made by placing the empty cuvette 11 on the support *10* and by introducing the compensator *12* in front of coil *5*. Then the cuvette is filled with the test oil and placed on the support *10*. Alternative magnetic field in the coil 4 is changed due to the presence of ferromagnetic wear particles as well as the output signal which depends on the magnetic permeability of the test oil. This signal comes to the preliminary amplifier 6, block of amplification and filtration 7, analog-to-digital converter *8*, and indicator *9*, where the particle quantity index PQ is shown.

The particle counter makes it possible to implement three techniques for direct measurement of the particle concentration. The first method involves measuring the concen-tration of the particles deposited on the RPD substrate. If visual observation of the wear particle type is not required, then an oil sample placed in a plastic cuvette is used in determining the particle type. This method is suitable for monitoring a mechanism when rapid direct determination of the wear regime is required.

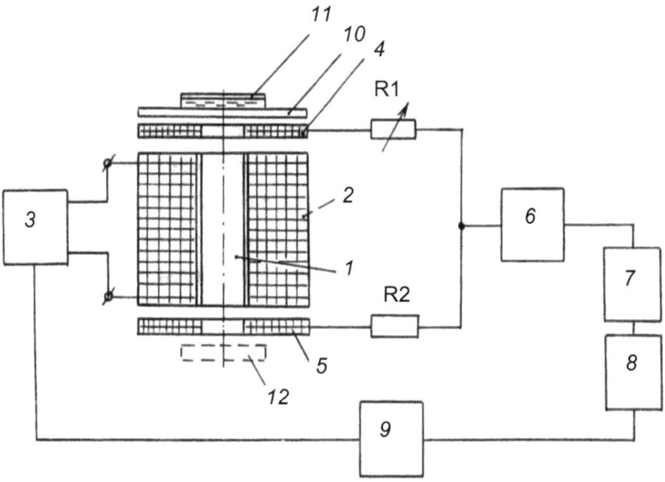

Fig. 5.7 Particle quantifier: *1* core, *2*, *4*, *5* electromagnet coils, *3* generator, *6* preamplifier, *7* block of amplification and filtration, *8* analog-to-digital converter, *9* indicator, *10* cuvette support, *11* cuvette, *12* compensator, *R1*, *R2* resistors [48]

The third method is used in the systems where the wear particle concentration is very small, such as hydraulic systems. In these cases a fixed large volume of the sample (e.g. 250 ml) is taken and passed through a 0.5 μm filter, it is flushed, dried, and subjected to measurement on the particle counter. The precipitate obtained on the filter can also be analyzed with the use of microscopy to obtain information on the particle size, shape, and composition.

The particle counter can be used under both laboratory and field conditions for rapid determination of the particle concentration in the oil. Its drawback is the influence of wear particle size on the counter indications.

Wear Particle Analyzer (WPA) was developed for estimating the presence of ferromagnetic particles in lubricants and hydraulic fluids by Tribometrics, Inc. [32]. It includes the special filter with ferromagnetic fibers (Fig. 5.8). The filter is installed inside the permanent magnet, the oil sample is pumped through the filter and the particles of wear are trapped in it. The accumulation of the ferromagnetic particles is changing the magnetic flux being registered by sensor. The data are related to concentration of particles which can be washed from the filter and studied with a microscope. WPA can be used for monitoring engines, transmissions, and hydraulic systems. It is applied in laboratories and installed at US Navy ships.

There are some other devices based on optical principles.

Devices of direct photometry are operated using the data on optical density related to the contamination of the lubricant by wear debris.

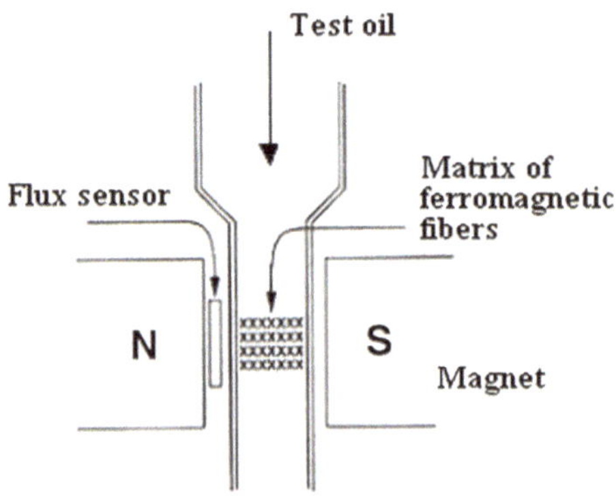

Fig. 5.8 Wear particle analyzer [32]

Light scattering technique is used for estimation of concentration of mechanical contaminants in oil basing on scattering of the optical flux in the test sample.

Flow ultramicroscopy provides the estimation of the concentration and size distribution of particles [49]. The test sample of liquid is coming through the capillary being in the focus of the optical source and the particles generate the flashes of scattered light being registered by the counter. This technique is used in the optical particle counters.

Basing on similar concept Lockheed Martin and the Naval Research Laboratory have developed a new device with the pulsed laser source LaserNetFines (LNF) which can classify and count the particles in the range of 4–100 μm [50]. The coherent pulsed laser light is transmitted through the oil and a silhouetted image of the particles recorded onto a digital charged coupled device (CCD) camera array (Fig. 5.9). Each particle pixel image is analyzed for the shape and structure of

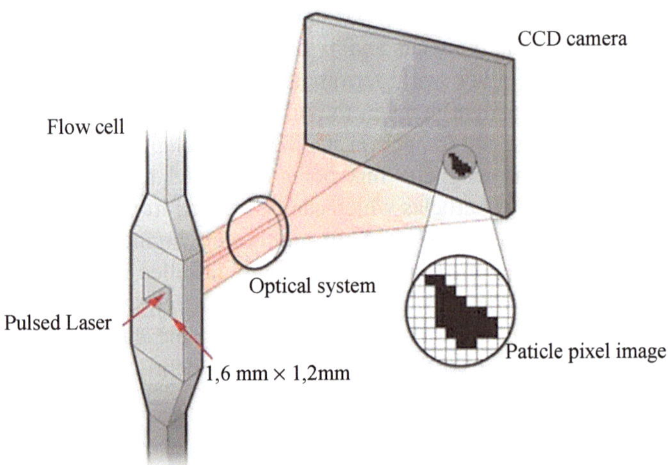

Fig. 5.9 LaserNetFines [50]

the particles by the image recognition techniques. The particles larger than 20 μm are classified into 6 categories based on the neural network approach: cutting debris, severe sliding ones, fatigue debris, non-metallic particles, fibers, and water droplets, while separating the air bubbles. LNF is successfully used in power generation and mining industries, but its main shortage is the problem of separation of the wear debris from the total contamination.

Portable devices cannot meet the current necessity to provide the data in a short time with the simple and reliable accuracy at the appropriate price. That makes the necessity to develop the real-time methods.

5.4 Techniques and Devices for Wear Diagnostics in Real-Time

In a general case they are characterized by sufficient information output at relatively low cost, simple mounting and maintenance, and high reliability. These methods are used in on-line and in-line devices and their diversity is presented in Fig. 5.10.

Methods based on wear debris accumulation use the effect of magnetic field produced by permanent magnet [51–53] or the electromagnet [54, 55] on ferrous wear debris deposit on magnet poles. Different methods are used to obtain data on

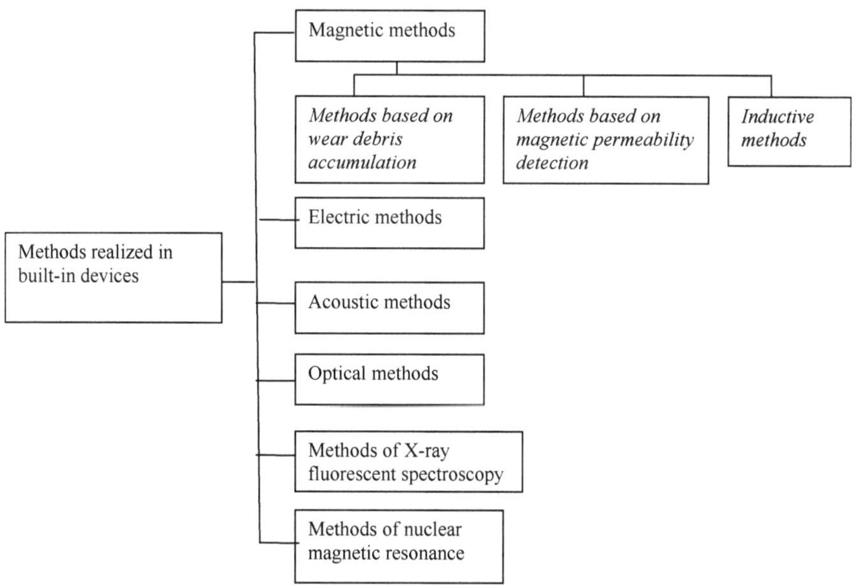

Fig. 5.10 Methods realized in built-in devices

the mass of accumulated particles. Thus, an additional measuring coil inducing electric current variation in the measuring circuit with ferromagnetic particle accumulation can be used [54, 55]. Capacitance of a capacitor with wear debris deposited between its plates is also used as an informative parameter in wear monitoring [56]. The analysis of accumulated wear particles is performed basing on measurement of optical signal by fiber optic system [57].

When the problem is to determine the maintenance necessity on the basis of the wear particle content in the oil, simple built-in devices based on particle accumulation can be used.

An example is the device containing a permanent magnet, a pair of installed pole pieces, and two temperature-matched magnetoresistors MR1, MR2, which form the arms of a Wheatstone bridge (Fig. 5.11a) [58]. A non-ferromagnetic spacer is installed in the oil line and the measuring device is mechanically connected with a motor that is used to displace it. When the device is brought near the

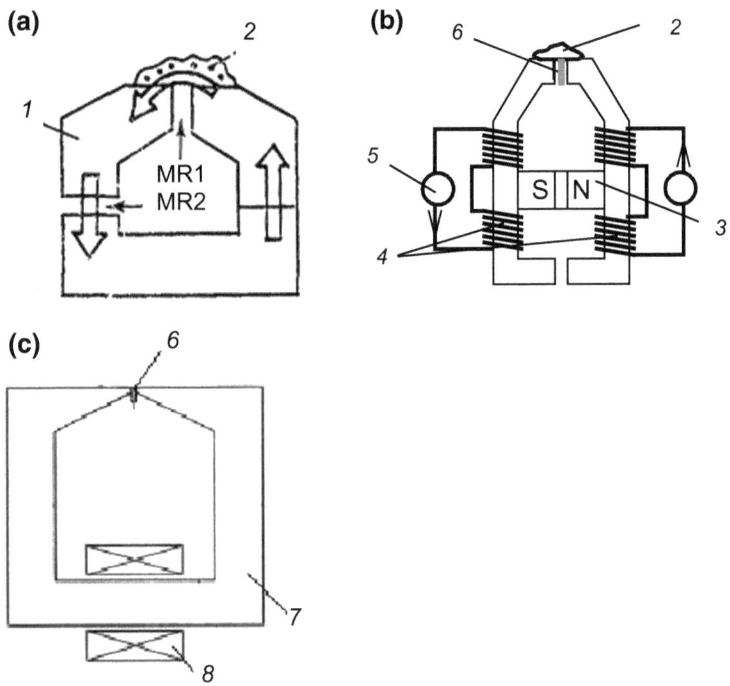

Fig. 5.11 Devices based on wear debris accumulation in gap of permanent magnet core (a) and electromagnet (b, c): *1* magnetic system; *2* accumulated wear particles; *MR* and *MR2* magnetoresistors; *3* permanent magnet; *4* induction coils; *5* current source; *6* Hall sensor; *7* magnet core; *8* electromagnet

non-ferromagnetic spacer wear particles accumulate on the unit. This causes change of the magnetic flux, which leads to reduction of the flux density in the magnetoresistive sensor MR1 and increase in the sensor MR2. The bridge output voltage will increase linearly with particle accumulation. Upon reaching a certain level the linearity is disrupted, the motor is energized and moves the metering device away from the spacer on which accumulation took place. The wear particles leave the accumulation surface. Then the measurement cycle repeats. This device can be used in relatively clean systems, in which the number of wear particles is critical. However, in measuring the particle content in the oil the particles whose size does not exceed 6 μm are not taken into account, since the magnetic flux gradient is not sufficient to capture them from the lubricant stream.

As an example of device using the method of magnetic accumulation we refer to the sensor for detecting the content of ferromagnetic particles applied by IVECO FIAT [59]. The sensor registers the wear debris accumulation in a magnetic circuit clearance (Fig. 5.11b). Variation in magnetic field induced in clearance 1 with increasing accumulation of wear debris 2 suspended in a fluid is recorded. The magnetic circuit including clearance 1 is polarized by permanent magnet 3. Induction coils 4 connected to the auxiliary magnetic circuit produce a magnetic field opposite to that produced by magnet 3. Electric current in coils increases up to disappearance of the magnetic field in clearance 1. As a result, wear debris accumulated above the clearance are separated, clean and regenerate the sensor.

Magnetic field of electromagnet can be also used for particle accumulation in the clearance of magnetic core 7 (Fig. 5.11c) [60]. Hall sensor 6 is installed in the clearance to monitor the accumulated particles. When the threshold level of Hall sensor output (which corresponds to limit level of mass of accumulated particles in the clearance) is reached, current in electromagnet 8 is switched off and the clearance cleaning occurs. This method is realized in Continuous Debris Monitor—CDM (Ranco Controls Ltd.).

The ferromagnetic detector [61] also realizes the method of magnetic accumulation, which records the variation in the generator reference frequency with wear debris accumulation in a magnetic trap (Fig. 5.12). A high-frequency source generates an electric signal at the main frequency and comprises sensitive coil 2 connected to it as an induction coil. Magnetic trap 1 produces the magnetic field. The sensitive coil and the magnetic coil are superimposed on each other and installed so that when the coil is mounted in a fluid and electrically connected to a generator the excitation of the magnetic trap results in attraction of ferromagnetic wear debris to the sensitive coil. As a result, the signal frequency deviates from its main value proportionally to the content of ferromagnetic material in a fluid.

Oil condition sensor based on measuring of resistivity variation caused by accumulated wear debris is described elsewhere [62]. The sensor is threaded into the wall 12 of an oil pan. The sensor (Fig. 5.13) head includes a ring-shaped permanent magnet 10 a cup-shaped common electrode 9 and eight rod-shaped

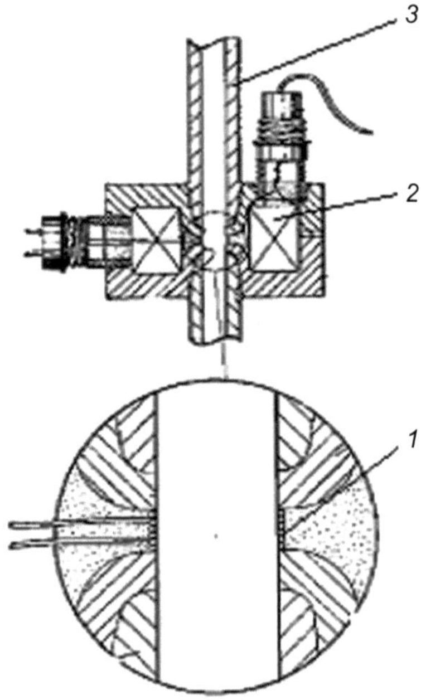

Fig. 5.12 Ferromagnetic detector: *1* sensitive coil; *2* magnetic trap; *3* pipe with test oil [61]

conductors 1–8 that are axially spaced from the common electrode through the insulator 11. The conductors are connected to a power supply through the separate resistors. Debris in the oil accumulates in the gap between the common electrode and the rod conductors because of magnetic attraction force. When the amount of metallic wear debris in the oil increases, a greater number of conductors are electrically connected to the common electrode. By counting the number of connected conductors, it is possible to estimate the oil contamination level by ferrous metallic debris.

There is a group of methods for estimating the presence of metal wear debris, as well as their concentration, and size. References [32, 63, 64] describe a built-in analysis device whose operation is based on accumulating the wear particles from the lubricant. The device contains a sensor that consists of an ensemble of permanent magnet 1 and the sensing electromagnet 2 (Fig. 5.14). Magnetic core 3 is dipped in the oil flow. Ferromagnetic wear debris are attracted to end of the core 3. This leads to change of the magnetic flux in the measuring coil, which causes the appearance at the output of a signal that is proportional to the magnetic flux rate of change, which is proportional to the mass of the captured particle. The signal from the sensor, which is a pulse whose amplitude is proportional to the particle mass, is fed through the preamplifier to the data processor, where the input pulses are separated into two categories, characterizing particles that are larger than 200 μm

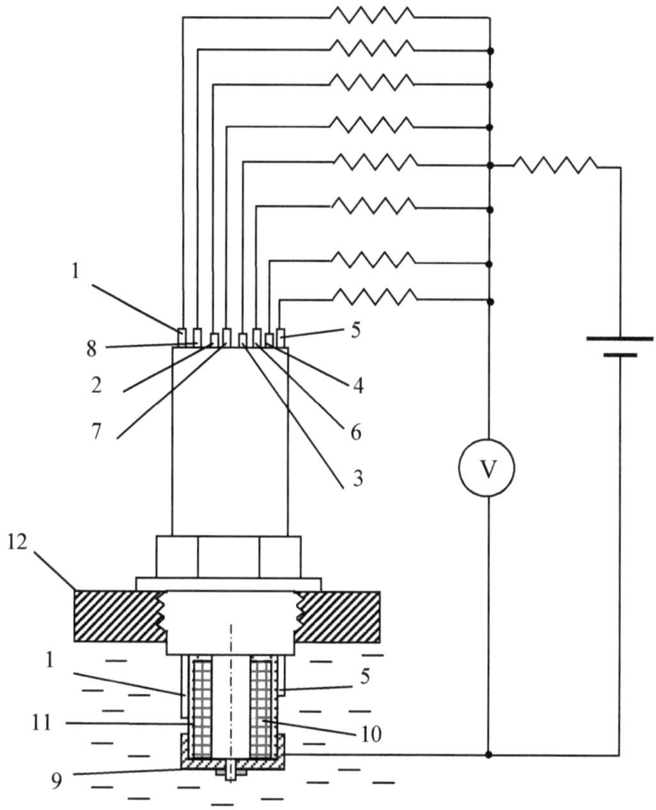

Fig. 5.13 Oil condition sensor based on measuring of resistivity: *1–8* eight conductors, *9* common electrode, *10* permanent magnet, *11* insulator [62]

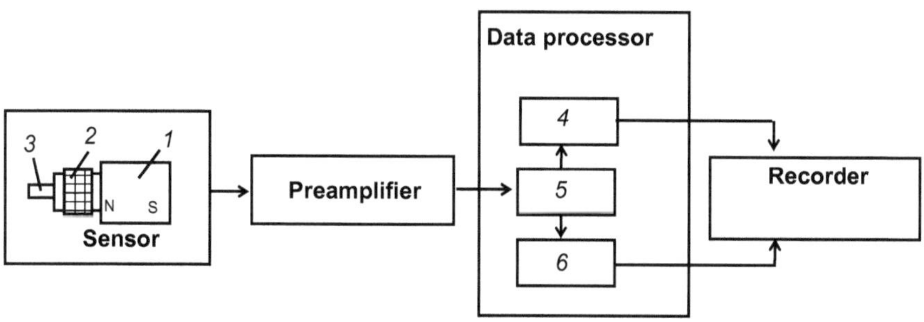

Fig. 5.14 System for qua ntitative wear particle monitoring: *1* permanent magnet; *2* measuring electromagnet; *3* magnetic conductor; *4* large particle counter; *5* pulse amplitude selection circuit; *6* fine particle counter

and particles that are larger than 1000 μm. After separation the pulses are applied to counters, which determine the pulse repetition rate and the number of pulses, and therefore the particle formation rate and the total number of particles. This method is realized in Quantitative Debris Monitor (QDM), produced by Eaton Engineered Sensors Co.

Simultaneous action of the electric and magnetic fields is employed in a chip detector [65]. It operates as follows. The test oil is passed through the electro-conductive magnetic electrodes forming the capacitor. With a certain amount of particles accumulated under the action of magnetic field the electrodes get closed and the capacitor is discharged. If just fine wear debris (<50 μm) have been accumulated they are dispersed and the capacitor is recharged automatically. Relatively large wear particles are not broken and in this case the severe wear indicator is turned on. This technique is operation basic of "Electric Chip Detectors" produced by Eaton Engineered Sensors Co.

Most informative sensor for built-in particle accumulation analysis is the device developed on the basis of direct reading ferrography where the optical system is used. Outputs of photoreceivers give information on density of accumulated particles on substrate [51].

The device consists of a direct reading ferrograph 10 (Fig. 5.15), two membrane pumps for the oil 5 and solvent 4, three cut-off valves 1, 2, 3 for controlling the flow, a signal converter 8, and a microprocessor 7, which controls the operation of the entire device and the output of information on the condition of the mechanism. The ferrograph itself is constructed with the use of the light source 11 and optical fibers, through which the information characterizing the number of fine and coarse particles is supplied to the signal converter, where the optical signal is converted with the aid of a. photocell to an electrical signal and enters the microprocessor for further processing. This system makes it possible to obtain information on the overall wear particle concentration and characterize the wear particle size distribution.

Modification of ferrographs is carried basing on state-of-the-art technology to adapt them to severe operation conditions, specifically, for monitoring of marine

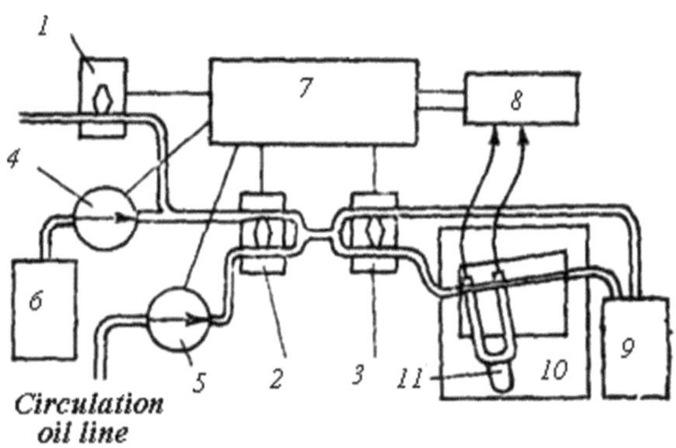

Fig. 5.15 Ferrograph built into the oil line: *1*, *2*, *3* cut-off valves; *4*, *5* membrane pumps; *6* solvent; *7* microprocessor; *8* signal converter; *9* drain; *10* direct-readout ferrograph; *11* light source [51]

diesel engines [66–69]. Wu et al. [67] developed an on-line visual ferrograph (OLVF) featuring automatic operation, visible debris image, and adjustable sampling parameters. Normally, an OLVF is connected to a lubrication system directly for on-line monitoring of wear debris.

Magnetic methods allow us to design the sensitive devices which can be used for registration of low wear debris concentration in light-loaded machines. But they need time for particle accumulation and flashing of the accumulated sediment which complicate the design of the devices.

The second group of magnetic methods is based on measuring the magnetic permeability of oil. Such devices commonly comprise two inductive coils, such as measuring and reference ones, the measuring coil encloses the pipe through which the oil is passed [70]. The occurrence of ferromagnetic particles results in the change in oil magnetic permeability and the unbalance signal is proportional to the mass of ferromagnetic substance found in the oil.

The methods based on the measurement of magnetic permeability are distinguished by simple and compact equipment and low cost. However they have a low sensitivity to small concentrations of wear debris

The third group of magnetic methods consists of techniques which are based on magnetic induction registration—inductive methods.

Such a method is realized in the Oil Debris Monitor (ODM, GasTop Ltd.) [71–75]. The monitor includes a magnetic coil assembly consists of three coils which surround a magnetically inert section of oil tubing (Fig. 5.16). The two exiting coils (1 and 1′) are driven by a high frequency alternating current source such that their respective fields are nominally opposed and cancel each other at a point inside the tube and just under the center sensing coil. As a result the output signal of sensing coil 2 is equal to zero. Disturbance of the magnetic fields caused by the passage of particle results in the characteristic sensing coil voltage. The amplitude and phase of the output signal is used to identify the size and type of particle. The amplitude is proportional to the mass of the particle for ferrous materials and to the surface area of the particle for non-ferrous material. The phase of the signal for non-ferrous particles is opposite to that of ferrous particles. Thus, analysis of output signal allows us to determine the number of non-ferrous and ferrous particles and to estimate particles size.

ODM is not accumulated the particles of wear, so it does not needs washing the sediment and can be used in severe operation conditions. ODM is used at fighters USAF F22 as a component of the engine Pratt &Whitney F119 as well as in the gear boxes of helicopters McDonnell Douglas AH64 Apache [32]. But its sensitivity is limited by large particles: ferromagnetic ones should be not smaller than 125 μm and non-ferromagnetic ones—250 μm [71].

Nowadays several research centers are searching a new design of the sensor to improve the sensitivity.

The in-line debris sensor based on dual excitation sources is described elsewhere [76]. The sensor includes measuring inductive coil, oil tube, permanent magnet, iron core, and housing (Fig. 5.17). The permanent magnet is used to generate static magnetic field. Iron core reduces reluctance of the magnetic circuit and increases

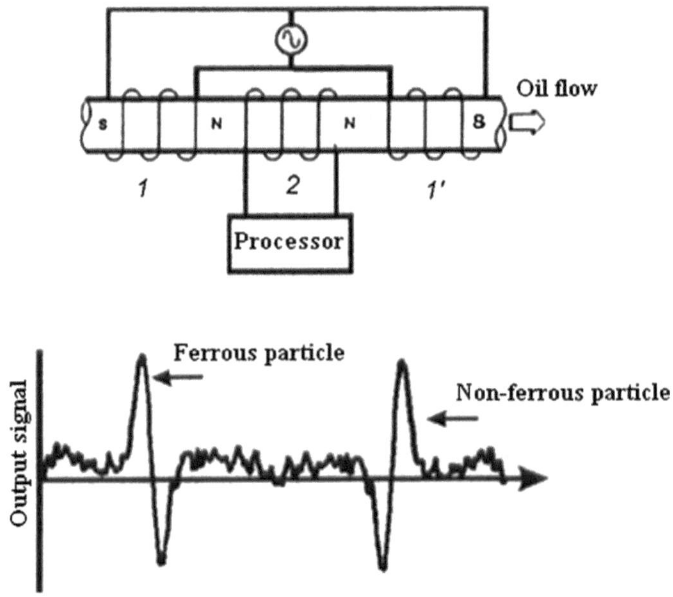

Fig. 5.16 Oil Debris Monitor (ODM): *1* и *1'* exciting coils, *2* measuring coil

Fig. 5.17 Sensor based on the dual excitation sources [76]

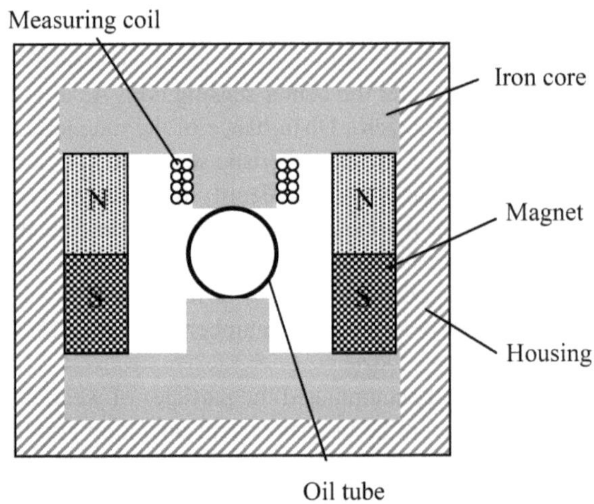

magnet strength in the oil tube zone. As a result the sensor's sensitivity is improved. The measuring coil records the change of magnetic flux caused by the debris. The housing is used to shield the sensor from electromagnetic interference from the surrounding environment. It was found that the sensor can detect 80 μm ferrous debris in 12 mm outside diameter of oil tube made of polymer.

In addition to magnetic methods, electric, acoustic and optical methods are used in design of novel built-in devices. The use of them for condition monitoring by wear debris is obviously restricted by machines whose lubricating systems are reliably protected against mechanical particles from the outside.

In most known devices based on the ***electric method*** of diagnostics a filtering element is used. Wear debris precipitate as the oil passes through the filter. This results in variations of the filter conductivity [77] or capacitance, which form the basis for evaluating the condition of the rubbing parts.

A microfluidic device (illustrated in Fig. 5.18) was developed to detect metal debris by monitoring the change in *capacitanc*e across a pair of microelectrodes in the microfluidic channel [78, 79]. The microfluidic device [79] includes an inlet oil tube, inlet reservoir 2, outlet reservoir 6 and outlet oil tube. The microchannel 4 represents a constriction between the two reservoirs. The height and width of the microchannel are selected to be sufficient to allow typical wear particles 3 that are to be detected to pass there through. For example, to register the particles of size about 40 μm the channel with dimensions of 40 μm × 100 μm × 400 μm is used. The reservoirs and microchannel are formed in the body 1 made of nonconductive material. The inlet oil tube includes the pump P to feed the test oil to the microchannel. The filter F is placed in inlet channel to avoid the blocking of microchannel by particles which exceed the minimum dimension of microchannel. A pair of electrodes 5 is disposed in the microchannel for detecting microparticles. When 10–100 μm metal particles suspended in lubricating oil passed through the microchannel, voltage pulses caused by change in capacitance between the electrodes are detected. Each pulse represents passage of metal debris. The voltage

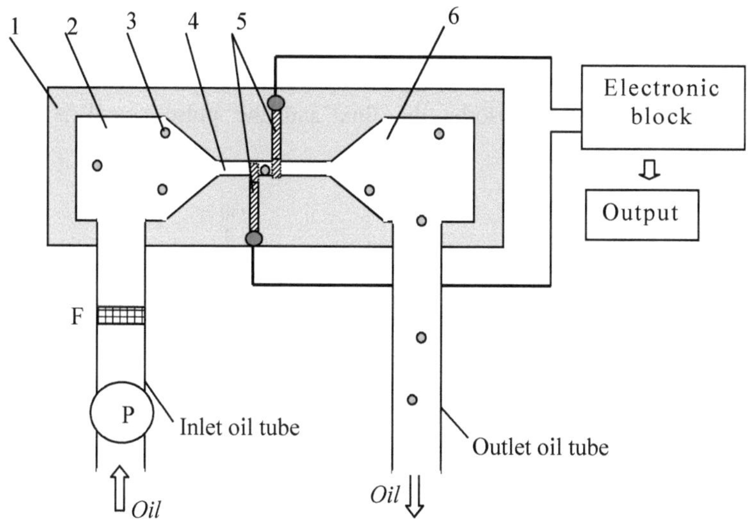

Fig. 5.18 Capacititance microfluidic device: *1* polymeric body, *2* inlet reservoir, *3* wear particle, *4* microchannel, *5* electrodes, *6* outlet reservoir, *F* filter, *P* pump [79]

pulses are processing in electronic block. Output signal relates to the number of wear particles.

While the device can detect very small metallic debris particles (~ 10 μm), the sensor is unable to distinguish the ferrous and nonferrous metal debris particles. Additionally, because the relative permittivity of water is rather high ($\varepsilon = 80$), it is difficult to distinguish the water droplets from metallic debris.

Acoustic method was realized in the particle detector [80] comprising a target which is bumped by solid particles thus giving rise to acoustic signals. Their energy is proportional to the particle kinetic energy. The acoustic signals are transformed to electric ones. After their processing an electronic block evaluates the total number and the mass of individual particles.

Another patent describes an ultrasonic debris sensor presented in Fig. 5.19 [81]. Two ultrasonic transducers are positioned on opposite sides of the oil pipe. One transducer is used as a transmitter sending ultrasonic wave, and the other transducer receives the ultrasonic wave. Attenuation of the ultrasonic wave amplitude in the receiver indicates the size and number of wear debris. The shortage of this device is that it is unable to distinguish the air bubbles from debris particles because air bubbles and wear debris particles cause similar acoustic attenuation.

The given method is difficult to realize in the detectors built into the lubrication system since the device is highly sensitive to acoustic noise of the system under operation and environment. The problem could be solved by filtering the useful signal from the acoustic noise, which is complicated.

Optical methods have been used also in design of built-in devices. Recent inventions use the advantages of fiber optics, such as miniaturization and decreasing electric noise.

Optical systems of particle counting based on scattering or absorbing the optical flux in the oil are widely used in the real-time monitoring devices [32]. In such a system the optical flux from the laser diode LD (Fig. 5.20a) is focused by lens *1* to the axis of the pipe *4* with oil flow. Optical system *2* transfers the flux to the photo receiver FD. Particle *3* absorbs the flux and the pulse signal is detecting

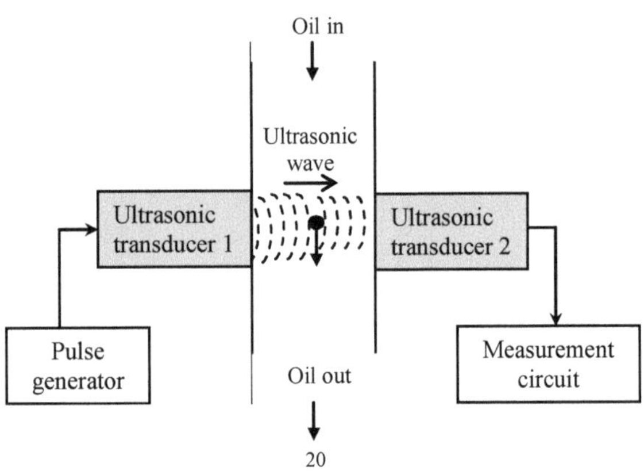

Fig. 5.19 Ultrasonic oil debris monitoring system [81]

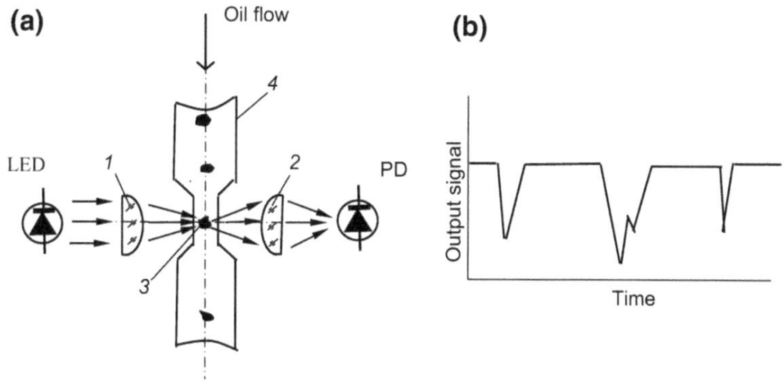

Fig. 5.20 Optical particle counter: *1* и *2* optical systems; *3* particle; *4* tube with test oil [32]

(Fig. 5.20b). Frequency and amplitudes of signals characterize the quantity and size of particles. Pacific Scientific Instruments produces such optical counter HIAC PM4000 [82].

As the optical counter can not distinguish the wear debris in all contaminants, as well air bubbles, and water drops there are methods using the magnetic field for identification of the ferromagnetic particles [83, 84]. The volume concentration of particles [83] or their mass concentration [84] can be found. For example the patent [83] describes the device where the test oil is pumped through the channel between the parts *7* and *8* (Fig. 5.21). Ferromagnetic particle is magnetized by coupled permanent magnets *1*. Hall sensor *2* mounted in the channel registers such particles and the optical system of diode *3*, windows *4*, lens *5* and photo receiver *6* detects all the particles, so correlating the signals of the optical system and Hall sensor it is possible to separate ferromagnetic debris in the total contamination particles.

It is possible to develop real-time methods on the base of *the X-ray fluorescent spectroscopy* used earlier in laboratory only [85]. In such a device (Fig. 5.22) the polymer chamber *6* is mounted in the oil channel 7 and the X-ray flux is focused at the area of I mm^2 in the channel via the window *3* from the X-ray source 1. This flux excites the fluorescence with the characteristic wave lengths of metals which is registered by detector 5 being Si-PIN photodiode. The output signal of the diode

Fig. 5.21 Optical detector of wear debris: *1* pair of permanent magnets; *2* Hall sensor; *3* light emitter diode; *4* optical windows; *5* lens system; *6* photoreceiver; *7* and *8* parts, forming the channel for oil flow [83]

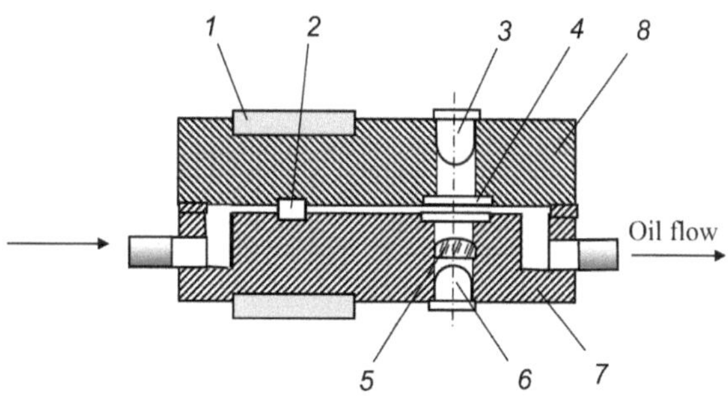

Fig. 5.22 X-ray fluorescence spectrometer for metal analysis in oil in real time: *1* X-ray source; *2* и *4* collimating/focusing units; *3* windows; *5* detector; *6* cell; *7* oil pipe [32]

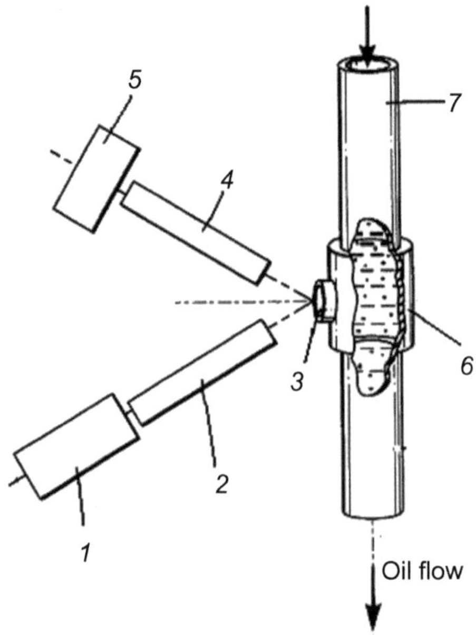

contains information on the quantity and type of the metal in the test oil and it is processed by computer. Optimization of the real-time methods is actively carried out by Pacific Northwest National Laboratory [86]. The company Outokumpu (Finland) in cooperation with the Allison Division of General Motors has developed the device for detecting titanium, iron, silver, and copper in oil which can be used at airplanes [32]. The analyzer 600T-LP of Spectro Analytical Instruments detect the elements with the atomic numbers from 12 (magnesium) to 92 (uranium).

Some research data are available on the use of the **nuclear magnetic resonance spectrometry** (NMR) in real-time monitoring systems [87]. Horseshoe permanent magnet *3* embraces the non-metallic pipe with the test oil *2* and the NMR indicator *1* (Fig. 5.23) which contains the coils of the alternative magnetic field excitation and the measurement coils. The device can detect metal particles and calculate their content in oil. The NMR disadvantage is a necessity to use large and heavy magnets for excitation of the strong magnetic field.

From the present review it is seen that most devices are based on magnetic methods, including numerous methods of magnetic accumulation and those measuring the oil magnetic permeability. It is caused by the fact that magnetic methods, when using in the heavy-loaded units made of ferrous materials, make it possible to identify wear debris among all the solid particles found in lubricant. The latter appear in the oil coming through the air filter, seals or gaskets from outside.

Besides, development of built-in devices is carried out basing on up-to-date optical methods, X-ray fluorescent spectroscopy and nuclear magnetic resonance spectrometry. The efficiency of diagnosis depends greatly on the type and size of wear debris. Mechanisms subjected to monitoring generate wear particles of

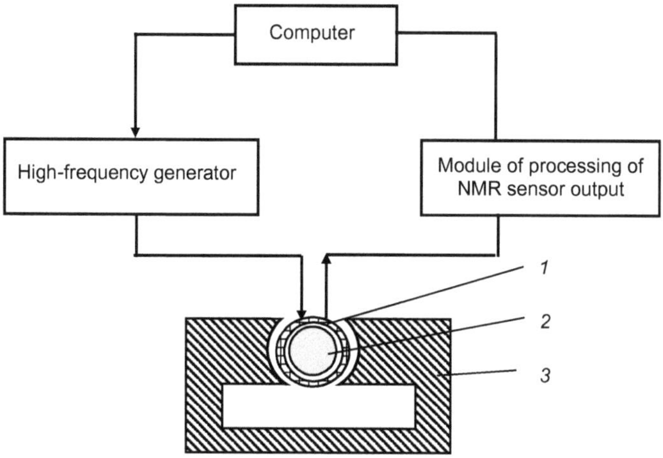

Fig. 5.23 Facility based on NMR for wear monitoring in real time: *1* NMR sensor; *2* pipe with test oil; *3* permanent magnet

different size, and a single particular method can not provide a high efficiency for all mechanisms. It should be chosen according to the preliminary experimental data on mechanism failure. In other cases several methods should be used.

The summary of the review of the optical and magnetic monitoring systems is presented in the Table 5.6.

The efficiency and reliability of built-in devices are governed by the complexity of oil system and their arrangement in the system which affect the wear debris concentration in oil in the zone of location of a sensor [88].

The Fig. 5.24 shows some schemes of possible mounting the sensor in the oil system. They are differed by the sensor position: it can be mounted in the main oil line (Fig. 5.24a) or in the by-pass one (Fig. 5.24b, c). The first type of mounting is characterized by a higher speed of analysis while the second type is more reliable. Such devices as MPD, QDM, CDM, Electric Chip Detector can be mounted in the main lubrication line (in-line devices), and QDM, Electric Chip Detector, HIAC PM 4000 and LaserNetFines can be mounted in the by-pass line (on-line devices). The drain of oil from the bypass line is provided either to the oil tank (Fig. 5.24b) or to the special drain tank (Fig. 5.24c, d). In monitoring the condition of individual units the detector is mounted in the accessory line of oil flowing from the unit being studied (Fig. 5.24d). If a particular unit should be identified, a similar scheme may be used with each unit being monitored. It should be noted that in developing on-line and in-line devices some problems have to be solved, such as the cleaning of the detector from debris and oil film, and wear diagnosis of the units operating with high-viscosity oils. These problems are partially solved when incorporating an additional tank (Fig. 5.24c) with a liquid for diluting the oil being examined or with solvent for cleaning the detector.

Table 5.6 Built-in devices manufactured nowadays

Device	Company	Operation principle	Type of detected particles	Particle size range, μm	Sensitivity
Ferrograph (Model 958PF)	Foxboro Analytical	Ferrography	Ferrous particles	1–100	–
Metal Particle Detector (MPD)	Smiths Industries	Monitoring of magnetic permeability variation	Ferrous particles	>100	2.0 mg (individual particles)
Quantitative Debris Monitor QDM	Eaton Engineered Sensors	Magnetic capture of individual particles and measurement of pulse amplitude and number	Ferrous particles	≥200	2.0 mg (individual particles)
Continuous Debris Monitor—CDM	Ranco Controls Ltd.	Magnetic accumulation and measurement of change in magnetic flow	Ferrous particles	>200	1.0 mg (collected deposition)
Electric Chip Detector	Aeton Engineered Sensors	Simultaneous actions of magnetic and electric fields on the oil under study	Ferrous particles	≥100	–
Oil Debris Monitor (ODM/MetalSCAN)	Canada; GasTOPS Ltd.	Registration of signal induced by particle in oil flow	Ferrous particles	≥100	–
			Metal particles	≥400	
On-line Liquid Particle Monitor HIAC PM4000	Pacific Scientific Instruments	Absorption of light by particle	All particles	4, 6, 11, 21	4 μg
On-line elemental analyze Spectro 600T–LP	Spectro Analytical Instruments Inc.	Energy dispersive X-ray florescence (EDXRF) technology	Concentration of elements	–	1% relative precision

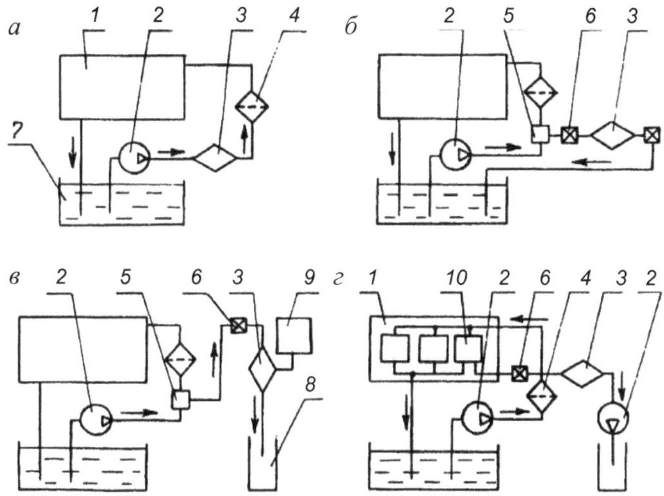

Fig. 5.24 The possible positions of sensor in the oil system: *1* test machine, *2* pump, *3* sensor, *4* filter, *5* pressure gauge, *6* valve, *7* main oil tank, *8* special drain tank, *9* diluents tank, *10* test unit

Therefore, in each particular case the selection of the optimal built-in device calls for preliminary determination of the type, size and concentration of wear debris formed in the operation of the mechanism being monitored, as well as of the scheme of mounting the detector into the lubrication line.

5.5 Development of Tribodiagnostic Devices

The efficiency of condition monitoring depends on the use of different devices because the systems under control (engines, transmissions, gears, etc.) generate wear particles in a wide range of sizes and shapes, so a single method and device cannot be efficient for all machines. The authors have developed a number of such methods and devices in Metal–Polymer Research Institute of National Academy of Sciences of Belarus. In some cases these methods were developed and tested in cooperation with the Korea Institute of Science and Technology [89–92].

5.5.1 On-line Opto-Magnetic Detector

The on-line Opto-Magnetic Detector (OMD) was developed for evaluation of tribosystem state basing on detection of oil degradation level and components wear.

Oil degradation is estimated by the total contamination of used oil which is defined by the variation in optical density of used and clean oil. Both contamination

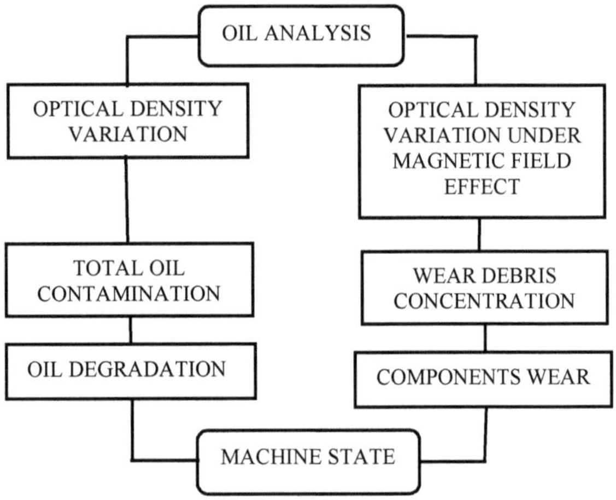

Fig. 5.25 Basis of evaluation of machine state with OMD

due to oxidation and ageing, as well as the presence of mechanical particles are considered, and the total contamination is used to estimate oil performance.

The variation in optical density of used oil under the effect of magnetic field is proportional to ferrous wear debris concentration in test oil, so the wear of machine components is defined by this concentration. The OMD allows one to assess the machine state on the basis of information on oil degradation and components wear (Fig. 5.25).

Theoretical basis of opto-magnetic method

The oil sample may contain wear particles in a wide dimensional range (0.001–100 μm), depending on the type of machine, loading conditions, and wear mode. Therefore, the oil sample exhibits both the properties of microscopic colloidal systems and the properties of macroscopic suspensions.

Upon application of a magnetic field on the oil sample the behavior of the wear particles will depend on their magnetic properties. It has been found that in the lubricants of iron-containing friction pairs there may be present both ferromagnetic iron components α-Fe, Fe_30_4, γ-Fe_20_3, Fe_3C and paramagnetic iron compounds $FeS0_4$, $FeS0_4 \cdot 7H_20$, $FeCl_2$, FeO, $FeC12 \cdot 4H_20$ [18].

In case of small size (≤ 0.1 μm) the ferromagnetic particles are individual domains, i.e., permanent dipoles. The particles of the paramagnetic iron compounds and large ferromagnetic particles have no permanent magnetic moment but acquire an induced moment in the external magnetic field. The magnitude of the induced magnetic moment $\overrightarrow{\mu}$ is determined by the volume V of the particle, its shape (particle demagnetizing field coefficient N_0), the magnetic susceptibility χ of the monolithic material, and the magnetic field intensity $\vec{H}$ [93]: $\overrightarrow{\mu} = \frac{V\chi\vec{H}}{1+N_0\chi}$.

This implies that even in the case of susceptibility that is relatively low but such that $\chi N_0 \gg 1$, $\vec{\mu}$ is independent of χ, and then $\vec{\mu} = \frac{V\vec{H}}{N_0}$. In the case of spherical particles (of radius a) $N_0 = \frac{4}{3}\pi$ and $\vec{\mu} = a^3\vec{H}$. The process of multi-domain particle magnetization results from shift of the domain boundaries within the particle. The dimension of the regions whose direction of magnetization coincides with the direction of the external field increases, while the dimension of the other domains decreases. If the particles are single-domain, then the primary role is that of the change of the direction of magnetization, i.e., the direction of its magnetic moment. Thus, the external magnetic field tends to align the magnetic moments along the field. Between the wear particles with magnetic moments $\vec{\mu}_i$ and $\vec{c}_j$ separated by the distance $\vec{r}_{ij}$, there arises dipole-dipole interaction, the energy of which is defined by the expression

$$U_{\mathrm{M}} = \frac{\vec{\mu}_i \vec{\mu}_j}{r_{ij}^3} \left[\cos(\theta_i - \theta_j) - 3\cos\theta_i \cos\theta_j\right],$$

where θ_i and θ_j—angles between $\vec{\mu}_i$ and $\vec{r}_{ij}$, $\vec{\mu}_j$ and $\vec{r}_{ij}$.

If we assume that both particles have the same moments $\vec{\mu}_i = \vec{\mu}_j = \vec{\mu}$ and are oriented along the intensity of the external field, i.e., $\theta_i = \theta_j = \theta$, then $U_{\mathrm{M}} = \frac{\mu^2}{r^3}(3\cos^2\theta - 1)$, where $r = \left|\vec{r}_{ij}\right|$.

This energy has a minimum for $\theta = 0$. Consequently, that mutual positioning of the particles is energetically more favorable if their moments are parallel and lie on a common straight line, passing through the centers of the dipoles. In case of relatively low concentration (<1 mass%) and the absence of other forms of interaction this leads to the formation of chains that are oriented along the field.

The behavior of the wear particles is determined by the relationship of the energies of dipole-dipole interaction, Brownian motion, interaction of the adsorption layers, and molecular interaction. The Brownian motion tends to maintain homogeneity of the system and its total energy includes the energies of rotational ($U_{br.r}$) and translational ($U_{br.t}$) motion [94]. The total energy can be represented as

$$U_{br} = U_{br,r} + U_{br,t} = \frac{(kT)^2 \rho_p}{\pi\eta^2} \left(\frac{1}{96a} + \frac{2n_p^{2/3} a}{27}\right),$$

where k—Boltzmann's constant; T—absolute temperature; η—dynamic viscosity of medium; n_p—number of particles per unit volume; ρ_p—density of particle.

The wear particles, interacting with the oil molecules, become covered with adsorption layers (of thickness δ), which prevents the coagulation of particles. In the process of Brownian motion and the action of external forces the particles can approach each other to a distance $s < 2\delta$ when the mechanism of steric repulsion of the particles is activated. We can evaluate the repulsion energy from the formula [95]

$$U_s = 8(kT)\pi a^2 N_{ad} \left[2 - \frac{l+2}{t} ln\left(\frac{1+t}{1+l/2}\right) - \frac{l}{t}\right],$$

where N_{ad}—surface concentration of adsorbed molecules; $t = \frac{\delta}{a}$, $l = \frac{s}{a}$, s—distance between surfaces of particles.

In addition Van der Waals attraction forces can act with magnitude depending on dielectric properties of particles and liquid. The interaction energy for two spherical particles of radius a with distance s between their surfaces can be found [96]:

$$U_V = -\frac{A}{12}\frac{a}{s},$$

where A—Hamacher's constant.

However, the molecular interaction energy is very small and it will be further neglected. In the case of uniform external magnetic field and the absence of other extreme actions in the case of small particles (<0.01 μm) the chain length is limited by steric repulsion and Brownian motion. A statistical theory of chains in the case of low dispersed phase concentration and a strong magnetic field was developed on the basis of the binary correlation function. It was shown that in the case of a strong magnetic field the particles form chains in which the average number of particles is evaluated from the formula

$$N_c = \left(1 - 2n_p|B|\right)^{-1}, \tag{5.1}$$

where B is second virial coefficient.

However, this theory is applicable only to small particles (<0.01 μm), for relatively large particles (~1 μm) even with low concentration (small n_p) the quantity $2n_p|B| > 1$ and formula (5.1) loses its meaning. In the case of large particles (>1 μm) in relatively strong fields sufficient for orientation of the magnetic dipoles the magnetic interaction significantly exceeds the Brownian motion. So, if the magnetic interaction energy exceeds the steric repulsion energy, there begins the formation of chains, the length of which is limited only by the cuvette dimensions.

Several works have been devoted to the experimental study of structure of ferromagnetic colloids basing on recording the change of the scattered or transmitted optical radiation in magnetic field, and also on observation of the formed chains by means of optical microscopy. The colloid being studied (average particle size <0.01 μm) was placed in a uniform magnetic field and the light passed through the sample in the direction of the magnetic field intensity $\vec{H}$. Upon activation of the magnetic field the intensity of light decreased more strongly, the higher the field intensity and the colloid concentration due to increase of radiation scattering by the formed chains. Chains of length ~0.9 μm were observed, which agrees with formula (5.1).

The study of the behavior of the wear particles in a nonuniform magnetic field is described elsewhere [93]. Such field had, on the one hand, an intensity component H_z along the z axis significantly exceeding the component H_r in the radial direction perpendicular to z. On the other hand, the intensity gradient along the r axis increases with increase of r and significantly exceeds the gradient $grad_r H$ along the

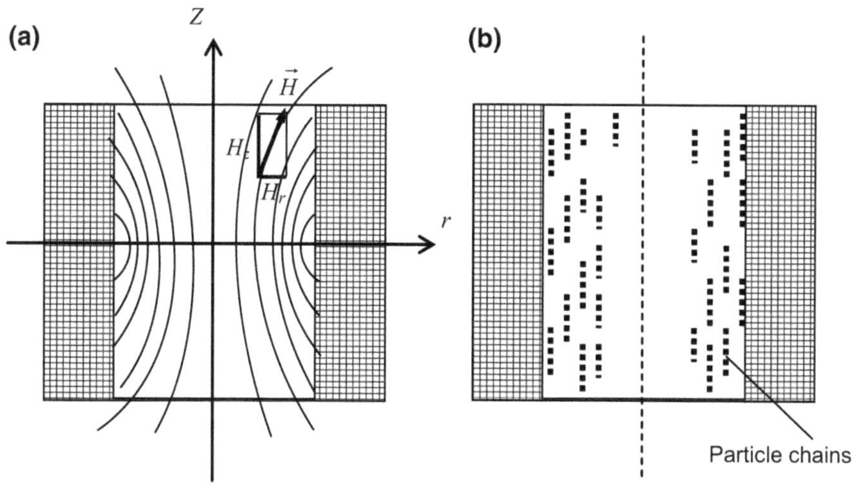

Fig. 5.26 Nonuniform magnetic field of solenoid (**a**) and ferrous particle behavior in the field (**b**)

z axis (Fig. 5.26a). To form such magnetic field the electromagnet with ratio 0.35 of inter diameter to length was used. In this magnetic field, just as in the uniform field, the chains are oriented along the z axis. However, in contrast with the uniform field case, the length of the chains of the large particles is limited by the magnitude of the gradient in the radial direction. From the condition of the energy balance of particle interaction with the chain and the work that is performed by the force that is caused by the gradient of the magnetic field intensity $grad_r H$ tending to displace the chain relative to the particle, we obtain the following relation for evaluating the number of particles in the chain:

$$\sum_{m=1}^{N_c} \frac{1}{m^3} \frac{Hn_p^{1/3}}{4\mathrm{grad}_r H} - N_c^{4/3} = \frac{8(kT)\pi N_{ad} n_p^{1/3}}{aH\mathrm{grad}_r H}. \tag{5.2}$$

In addition the magnetic field gradient in the radial direction leads to the presence of a force that tends to shift the chain from the axial zone towards the periphery. Taking into consideration the force of repulsion between the chains (taken as dipoles) and the force acting on the chain that is caused by the nonuniformity of the field, we obtain the formula for determining the distance between the chains:

$$L = \left(\frac{a^3 H_z N_c}{\mathrm{grad}_r H} \right)^{1/4}, \tag{5.3}$$

where H_z—projection of vector of external magnetic field intensity on the axes z.

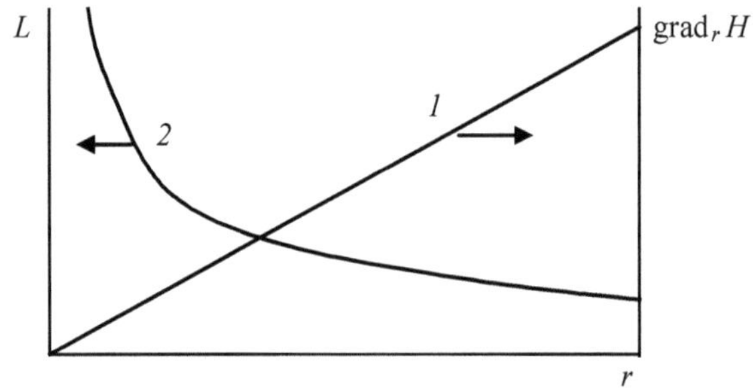

Fig. 5.27 Change of the gradient of the magnetic field intensity *grad*$_r$*H* (*curve 1*) and distance between chains (*curve 2*) in radial direction *r*

Radial gradient of solenoid magnetic field increases when r increases. As a result density of the chains is minimal in the axial zone and maximal at the periphery (Fig. 5.27).

The relations (2) and (3) were supported by the test data and for particle size 0.5–12 μm and chains of length 300–700 μm it was found N_c = 100–300 with change of the density of the chains along the *r* coordinate.

Thus, when a cell with oil containing ferro-magnetic particles is placed in a magnetic field with special configuration the particles form chains and can be shifted from the cell axis (Fig. 5.26b). If an optical beam passes along the axis the value of optical emission in the cell outlet varies before and after application of a magnetic field, i.e. optical density of the oil varies under the effect of the field.

In general, an optical density of the oil can be evaluated as follows

$$D = \ln(I_{01}/I_{02}) = k_\lambda l, \tag{5.4}$$

where I_{01}, I_{02}—intensities of incident and transmitted radiation, correspondingly, k_λ—absorption coefficient of medium for wavelength equal λ, *l*—length of optical path in medium. Taking into account relationships $I = \Phi/A_s$ and $U = \Phi S$ formula (5.4) can be written as

$$D = \ln(U_{01}/U_{02}),$$

where Φ—flux of optical radiation, A_s—area of irradiated surface, *U*—output voltage of photoreceiver, *S*—integrated sensitivity of photoreceiver.

The variation in optical density of clean and test oil can be estimated by formula:

$$\Delta D_1 = D_2 - D_1 = \ln\frac{U'}{U_1} - \ln\frac{U'}{U_0} = \ln\frac{U_0}{U_1},$$

where D_1 and D_2—optical density of fresh and used oil, correspondingly, U'—electrical signal when the cell is empty, U_0—output signal of cell with fresh oil, U_1—output signal of cell with used oil. The variation in optical density of test oil in magnetic field is evaluated by the formula:

$$\Delta D_2 = \ln(U_2/U_1),$$

where U_2—output voltage when magnetic field affects the oil.

Portable opto-magnetic detector

Based on the results obtained in theoretical and experimental studies of debris behavior in oil under the effect of magnetic field, the opto-magnetic technique and detector were developed in MPRI. This detector (OMD) consists of mechanical and electron blocks (Fig. 5.28).

The mechanical block includes light source 7, photoreceiver 6 with a preamplifier, optical transmitters 2, 5, magnetic system 3, and cell with oil sample 4. The light source and photoreceiver were chosen based on the analysis of the light-transmission spectrum of lubricant, spectra of source and receiver. The spectra of pure and used (6000 km run) oil M-8B were taken from the infrared spectrophotometer UR-20. Spectrum analysis shows that there are three explicit absorption bands having maxima at wavelengths 3.5; 6.7; 7.3 μm, and maximum

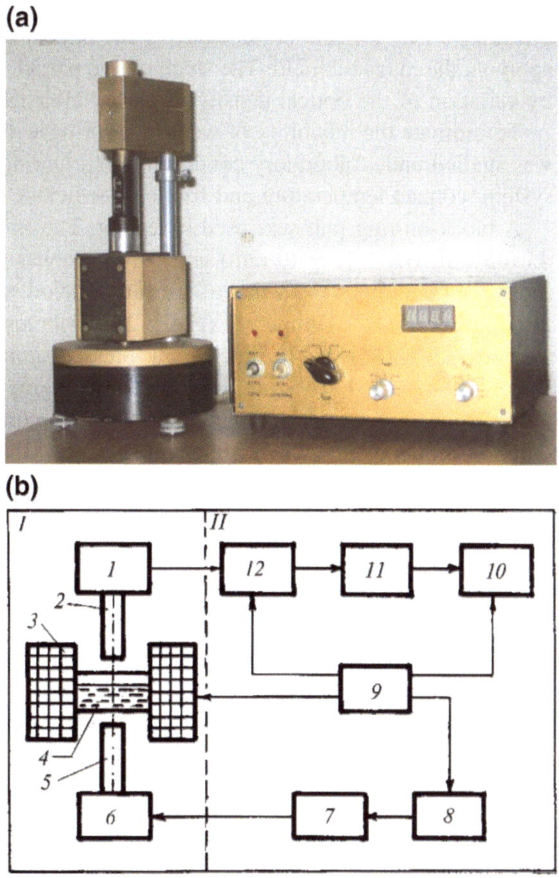

Fig. 5.28 Photo (**a**) and block-diagram (**b**) of opto-magnetic detector: *I*, *II* mechanical and electron blocks; *1*, *6* source and receiver of light; 2, *5* optical system; *3* magnetic system; *4* oil container; *7* radiating power controller; *8* pulse shaper; *9* power unit; *10* recording device; *11* filter; *12* amplitude detector

bandwidth over intervals 1–3, 4–6, and 8–13 μm. The band 1–3 μm is the most suitable for operation taking in mind the ease of operation of light sources and receivers. To ensure the passage of light through the "dark" (dirty) oils, an IR diode is chosen, its maximal radiating power being at a wavelength of 1 μm. A photodiode PD-27 is used with spectrum sensitivity maximum at the same wavelength. The magnetic system produces the nonuniform magnetic field affecting the oil sample. The electron block consists of the amplitude detector 12, recording device 10, pulse shaper 8, radiating power controller 7 and power unit 9. The pulse shaper converts the power unit voltage into pulses of frequency 2 kHz, relative pulse duration 20–30%, and amplitude 2 V, which makes it possible to obtain the pulses of voltage 20 mWt at the IR diode output. Before feeding the pulses to the diode they pass the controller varying pulse amplitude. Light passing through the container is converted to the pulse electric signal and comes to the preamplifier which amplification coefficient from 1 to 200. The amplified signal passes through the amplitude detector and filter for rectifying the signal. Then the signal is applied to the recording device, viz. the analog-to-digital converter with digital indication.

The total oil contamination is estimated by varying optical density of the oil sample relative to pure oil density. Concentration of ferromagnetic wear debris is determined using a calibration chart by the optical density of the oil sample when applying the magnetic field. The share of fine particles suspended in the oil is found by variation of the optical density of the oil after taking-off the magnetic field.

To estimate the reliability of wear monitoring with OMD the tribosystem run-in was studied under laboratory conditions [97]. During testing the linear wear of the system, contact temperature and friction coefficient were recorded.

A block-on-ring pair was used in testing. The ring was made of hardened steel 45 (45…50 HRC, R = 40 mm) and the block was made of unhardened steel 45. Industrial oil I-20A was used as a lubricant. The oil was poured into bath 3 in which the ring was partially immersed (Fig. 5.29). Tribotests were made with the friction machine 2070 CMT-1. The block 1 was pressed against ring 2 with force controlled by loading device 4. The ring was rotated by engine 9. Sliding velocity was set by varying the engine revolutions number and recording the value with velocity detector 8. In testing, the sliding velocity was kept constant and amounted to 0.5 m/sec, the pressure varied stepwise: 1, 2, 3 MPa.

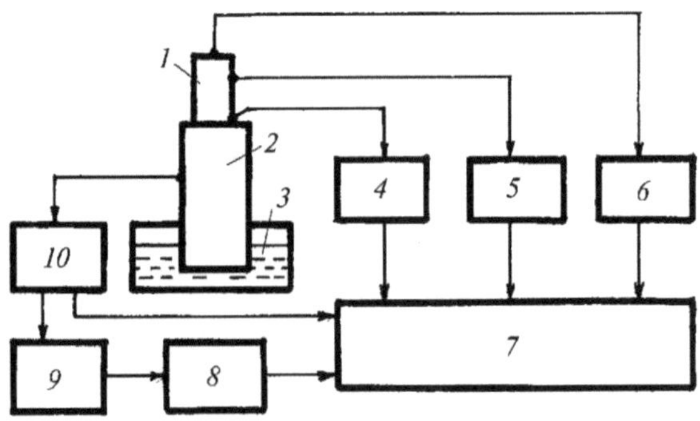

Fig. 5.29 Block-diagram of the testing complex: *1* conformable block; *2* ring; *3* oil bath; *4, 5* temperature and wear detectors; *6* loading device; *7* control unit; *8* velocity detector; *9* engine

At fixed moments of time the machine was stopped and after stirring the oil, the 2.5 ml sample was taken. Then pure oil was added to the bath so that the initial volume was reached 75 ml and the test was continued. The oil samples were examined by OMD. At high wear debris concentration the oil sample was diluted with the pure oil in the ratio 1:1. The test results indicate (Fig. 5.30) that the stepwise pressure variation results in change in the state of the lubricant film. In particular, its failure leads to seizure, increase in friction coefficient, temperature, and wear debris generation rate. The linear wear is proved to correlate well with the wear debris concentration. Contact temperature and friction force are correlated with the rate of debris content variation. Inflection points of all dependences occurred in stepwise loading, coincided in time and corresponded to the run-in start. An increase in temperature and friction coefficient coincided with the increasing rate of debris content which corresponded to an increase in contact pressure. The results indicate that OMD monitors successfully the wear of the friction pair.

The device performance was compared to that of the particle quantifier (PQ) developed in the Tribology Center of Swansea University College [48]. For measurements, the model oil samples with ferromagnetic particles of carbonyl iron of size 0.5…0.12 μm and concentration 0.005; 0.001; 0.05; 0.1; 0.2% by mass were used. The samples were made by weighing the iron powder on a torsion balance and further diluting it with 100 ml of oil. Two types of oil were used—industrial oil

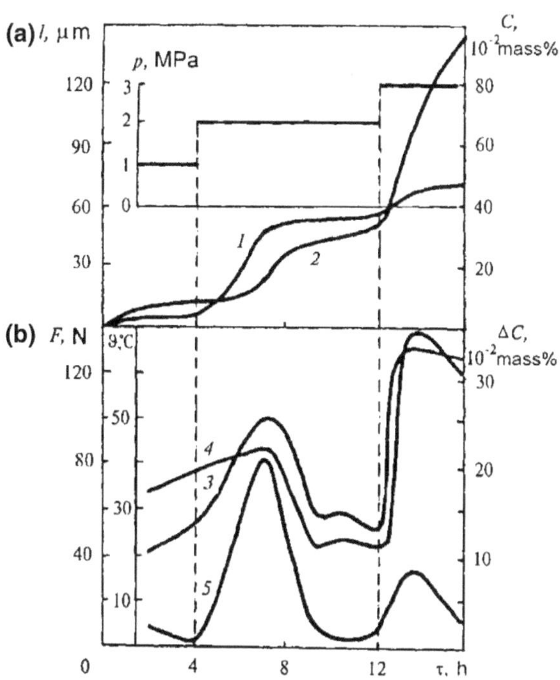

Fig. 5.30 Time dependencies: **a** wear debris concentration in the oil sampling (*1*) and linear wear (*2*); **b** variation of wear debris concentration in the oil sample (*3*), temperature in the friction zone (*4*), and friction force (*5*)

Table 5.7 Results of model sample analysis

No. of sample	Ferromagnetic particles concentration, % by mass	ΔD	PQ index
	Industrial oil I-20A		
1	0.01	0.0089	20
2	0.05	0.1704	140
3	0.10	0.4855	270
4	0.20	1.0048	570
Aviation oil MC-20			
5	0.005	0.0017	10
6	0.010	0.106	20
7	0.050	0.1973	120
8	0.100	0.4900	200
9	0.200	1.0349	490

I-20A and aviation oil MC-20 having different optical density. The comparative analysis was based on determining PQ index by the particle quantifier and variation of optical density ΔD by OMD for the model oil samples. In determining PQ index an average value was taken after six measurements. To analyze the oil by PQ and OMD the sample of 1 ml was taken. The test results show (Table 5.7, Fig. 5.31) that both devices are efficient over the whole particle concentration range. It is shown by analyzing oil samples taken from the lubrication systems of real transmissions (Table 5.8).

Thus, by its characteristics OMD is highly competitive with analogs and makes it possible to monitor both oil performance and tribosystem wear. It provides the determination of total contamination of the used oil, ferromagnetic wear debris concentration, and percentage of fine particles. The minimal ferromagnetic particle concentration to be revealed is 10^{-3}% by mass, oil sample volume is 0.5–2.0 ml.

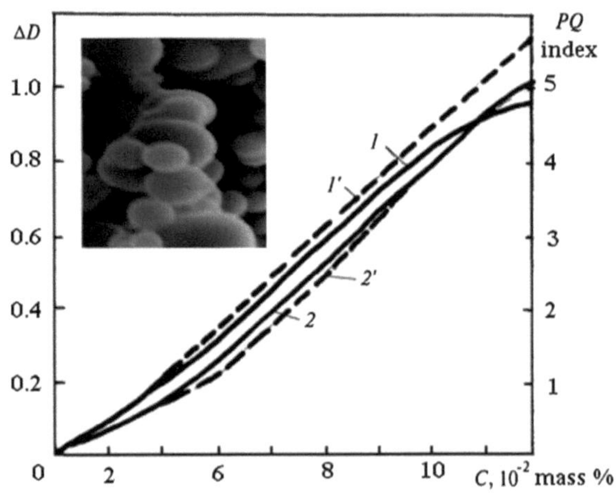

Fig. 5.31 Test results of model oil samples: 1, 1′ *PQ* index (particle quantifier); 2, 2′ ***ΔD*** (OMD); 1, 2 oil I-20A; 1′, 2′ MC-20. In the setting-in there are shown carbonyl iron particles, x10[4])

Table 5.8 Results of real sample analysis

No. of sample	OMD		*PQ* index	Conclusion based on analyzing the debris deposit
	Wear debris concentration, $10^{-2\%}$ by mass			
	Total	<50 μm		
1	0.7	0.65	20	Medium sediment density: mainly particles 15 μm in size are available; particles of 40 μm are found
2	3.0	2.50	80	Relatively high sediment density: mainly particles 20 μm in size are available; particles 80 μm in size are found
3	0.8	0	20	Medium sediment density: most particles are 65–130 μm in size. Copper particles 10 μm are available; high wear rate

On-line OMD System

One of the main problems of oil testing in real lubricating system is the high content of bubbles in oil. In particular, optical sensors of built-in optical device are sensitive to air bubbles, as well as to wear particles. This leads to erroneous diagnosis results. Therefore, the problem of air bubble removal from the test oil and wear particles recognition among particles and bubbles should be solved for successful operation of built-in optical detectors. Among existing methods of removing air from oil vacuuming is more widely used in engineering and this method is used in on-line OMD system developed by MPRI and KIST [89, 90].

The on-line system (Fig. 5.32) contains OMD sensor *1* and its electronic block 2, analog–to–digital converter (ADC) 3 and I/O control unit 4 by means of which acquisition and processing data and controlling procedure are carried out by computer 5. Test oil from tank 7 of oil system is carried by vacuum pump 6 to OMD sensor.

The measurement technique includes following procedures:

- filling system by oil—oil flow under vacuum pump 6 action at opened valves 8, 9 passes from the tank 7 with test oil to the OMD sensor 1;
- vacuum processing runs at closed valves 9 and 10 under given pressure and duration dependent on oil viscosity;
- testing oil—the first reading $U_1[i]$, $i = 1...50$ is taken, then magnetic system turns on and the second reading $U_2[i]$ ($i = 51...150$) is taken;
- draining oil from system goes at open valves 9 and 10.

The condition monitoring with OMD system is performed with the following indexes: total oil contamination index C_t and wear rate index W_r. Index C_t is estimated by formula: $C_t = ln(U_0/U_1)$, where $U_1 = \sum_{i=1}^{50} U[i]/50$, U_0—output at clean test oil.

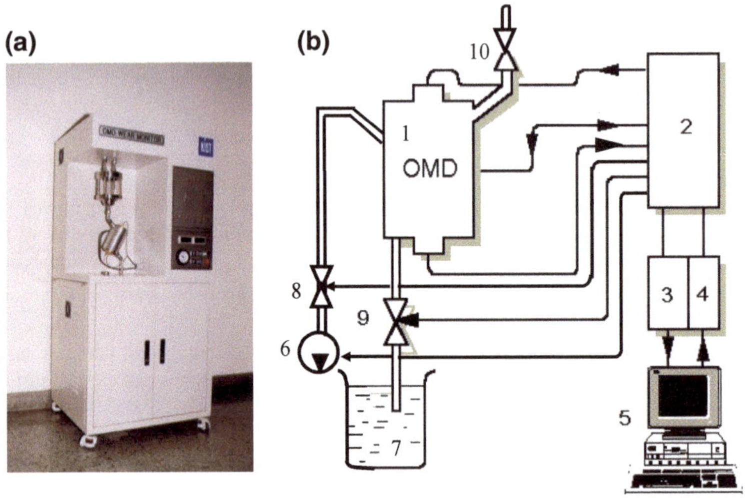

Fig. 5.32 Photo (**a**) and scheme of on-line OMD system: *1* OMD sensor; *2* electronic block of OMD; *3* ADC; *4* I/O control unit; *5* computer; *6* vacuum pump; *7* oil tank; *8*, *9*, *10* electromagnetic valves

Wear rate index is determined as: $W_r = ln(U_2/U_1)$. The values C_t and W_r are compared with limiting values C_{tmax} and W_{rmax} for machine under monitoring. A criteria of normal machine condition is $C_t < C_{tmax}$ and $W_r < W_{rmax}$.

Laboratory Test of the On-line OMD

The scheme of laboratory set was analogous to one presented in Fig. 5.32. The vessel with test oil was used instead of real unit 7. In order to imitate the real condition the mixer with high speed was mounted in this vessel to prevent particle deposition and to produce air bubbles in oil.

Laboratory test of on-line OMD was performed in Korea Institute of Science and Technology with unused compressor oil containing standard contaminants [91]:

- carbonyl Iron Powder (CIP) with concentrations (C): 5, 10, 20, 30, 40, 50, 70, 100 and 200 ppm;
- ISO standard contaminants of Air Cleaner Medium Test Dust (ACMTD) with concentrations: 5, 10, 20, 50, 100 and 200 ppm;
- combined contaminant 50% CIP + 50% ACMTD with total concentrations: 10, 20, 50, 100 and 200 ppm.

Vacuum processing oil was performed at pressure 74 cm Hg during 1 min. The measurements were carried out three times for each concentration. Results of tests are presented in Figs. 5.33 and 5.34 as the dependencies of variations in optical density on particle concentration. These dependencies have a linear character for each tested contaminant. Some deviations can be attributed to the error at of oil sample preparation at low concentrations specially. Differences in slopes of

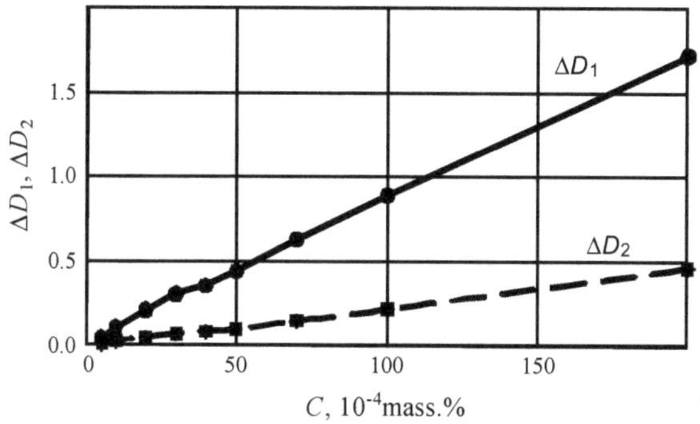

Fig. 5.33 Relations between optical densities ΔD_1 and ΔD_2 and concentration of CIP contaminant in test oil

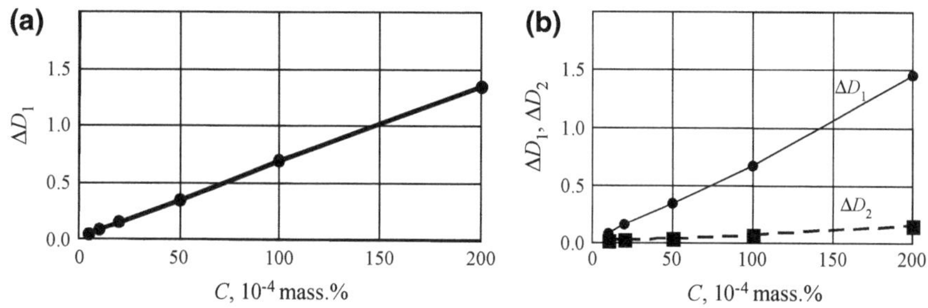

Fig. 5.34 Optical density change with ACMTD contamination (**a**) and with 50% CIP + 50% ACMTD contamination (**b**)

$\Delta D_1(C)$ for oil with CIP, ACMTD and CIP + ACMTD contaminants may be due to different particle size distributions of contaminants CIP and ACMTD.

Difference in slopes of $\Delta D_2(C)$ in figures in cases for CIP and CIP + ACMTD contaminants is related, probably, to screening effect of ACMTD which reduces magnetic interaction between ferromagnetic particles.

The results of laboratory tests show that OMD has a simple design and linear relation between variations in optical density and concentrations of particles in oil in a wide range (10–200 ppm).

Application of on-line OMD system for air compressor monitoring

On-line OMD was used for monitoring the air-turbo compressor system during two years. In steel making industry this system needs a secure condition monitoring because of its importance. The maintenance history showed that this system was regularly overhauled every two years, and irregularly serviced several times when the abnormal performance occurred. Main objects monitored were gear unit and metal bearings, both of which work under severe operation conditions. Vibration and temperature monitoring sensors were used in this system and positions of

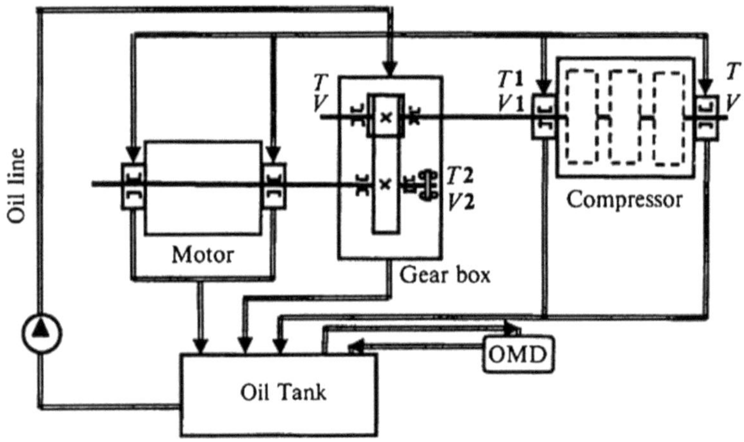

Fig. 5.35 Schematic view of air-compressor system. *V*, *T* vibration and temperature sensors

vibration and temperature detectors are shown in Fig. 5.35. OMD was installed in bypass line.

Vibration sensors measure the axial displacement at the air compressor input axis (VI) and acceleration at the gear box unit (V2). The temperatures were measured on metal housings of compressor input axis bearing (Tl) and gear box bearing (T2).

In addition, off-line wear monitoring with various types of detectors was used in the course of performance of test on-line OMD. The content of ferrous particles was quantitatively estimated by a Particle Quantifier (PQ) and Inductively Coupled Plasma (ICP) spectroscopy. The content of all solid particles in the oil was measured by a digital Contam-Alert (dCA).

Figures 5.36 and 5.37 present results of monitoring for thirty days. We can clearly see that OMD gives an early warning sign according to both indexes (total oil contamination C_t and wear rate index W_r) 10 days earlier than the system failure.

The measurement results were summarized in the Table 5.9. Increase in PQ index and Fe content in ICP analysis prove clearly that the wear of ferrous machine components took place.

Simultaneous increase in number of all solid particles indicates that the wear of ferrous component is the main reason of wear and variations in total oil contamination. OMD data have close agreement with ICP, PQ and dCA measurements.

These results contrasted to vibration and temperature measurements (Figs. 5.36 and 5.37). Readings of vibration and temperature sensors show a slight increase in signals just before the system failure, but they were far below the warning level.

Air-compressor system was shutdown only when vibration signal was abruptly increased. After the system shutdown, it was found that the collar (30CrMoV9 steel alloy) preventing the slippage of pinion gear in the axial direction was severely broken. The collar revealed severe sliding marks on the surface facing the wheel

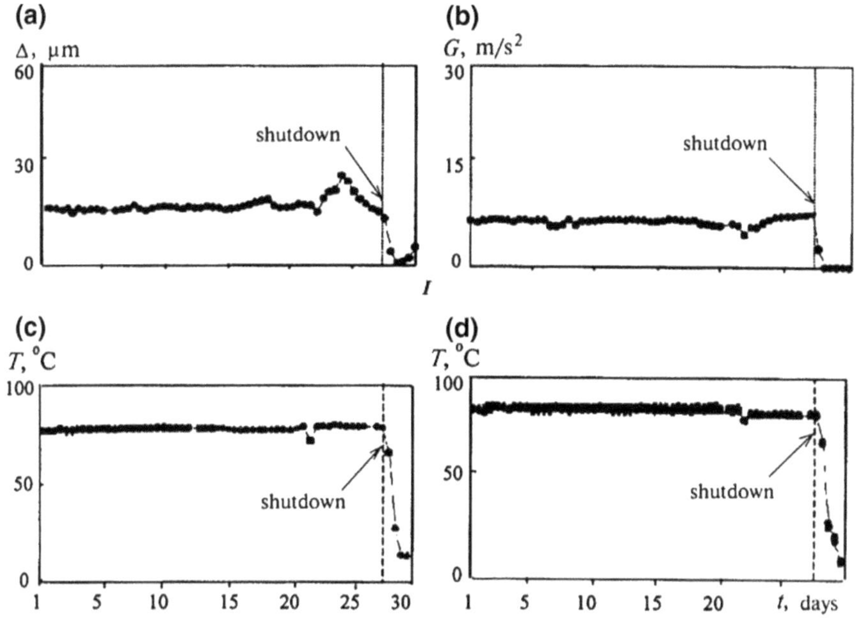

Fig. 5.36 Variations of test parameters in compressor operation: displacement of compressor driving axis (**a**) and acceleration in the gear box (**b**); temperature in the axial bearing of compressor drive (**c**) and gear box bearing (**d**)

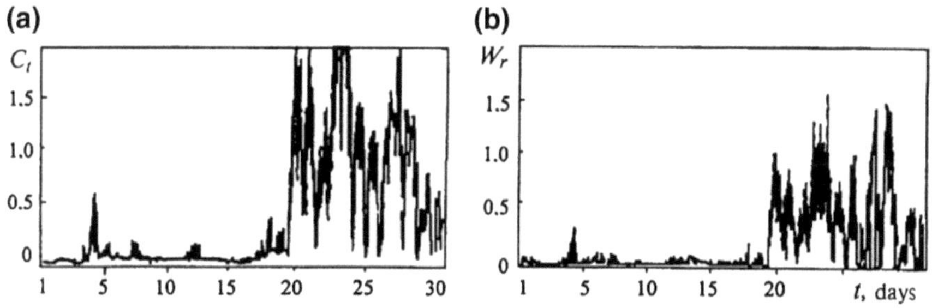

Fig. 5.37 OMD readings: the variations in total contamination index (**a**) and wear rate index (**b**) with the running time

Table 5.9 Comparison of off-line and OMD results

Time, days	ICP	PQ	dCA	OMD	
	Fe (ppm)	PQ Index	No. of particles	C_t Index	W_r Index
1	1.0	9	30	0.10	0
8	2.2	6	36	0.15	0.05
22	30.0	10	72	1.30	1.00
27		90	310	2.00	1.50

gear, so there was a direct contact with the gear. When the system failure occurred, it was stopped for more than 10 days and the accident resulted in the large-scale loss.

The on-line Opto-Magnetic Detector (OMD) can be an effective tool in condition monitoring. The detector offers an instrument capable of estimating the oil degradation and severity of wear occurring in mechanical equipment in real time. OMD has a simple design and linear relation between variations in total contamination and wear indexes and concentrations of particles in oil in a wide range (10–200 ppm). It was found that there are applications where OMD is more sensitive than other monitoring techniques in particular vibration and temperature control.

Obviously that in condition monitoring of critical equipment the best way is to use several types of techniques in order to compensate the limitations of particular ones.

5.5.2 Optical Ferroanalyzer

The method of DR Ferrography has been improved in Optical Ferroanalyzer (OF) jointly developed by V.A. Belyi Metal-Polymer Research Institute of National Academy of Sciences of Belarus and Korea Institute of Science and Technology. Optical Ferroanalyzer in addition to Ferrograph can estimate total contamination of oil, increasing reliability of condition monitoring.

The analyzer principle is based on the analysis of the concentration of ferrous particles in lubricant upon the changes in the optical radiation reflected by the substrate in the zones of accumulation of large and fine particles under the effect of the magnetic field and by rating the lubricity of the oil when registering its optical density changes [92].

Figure 5.38 shows the optical ferroanalizer sensor. The oil from the friction unit passes through a channel and electromagnet 5 induces the field in some part of the channel. The field is very strong with a high gradient in the direction perpendicular to the oil flow, ferromagnetic particles precipitate thus in two measuring zones of the channel (*1*, *1′* and *2*, *2′*). The particles are analyzed by measuring their concentration in the lubricant based on changes in the light reflected by the substrate 4 in the zones of accumulation of large (*1*, *1′*) and fine (*2*, *2′*) particles. The emission of the sources 1 and 2 is transmitted by optical fibers to the accumulation zones where it is reflected by the substrate 4 and passed through the optical fibers 9 to photodetectors (*1′*, *2′*), that transform it into electric signals. The optical fibers allow one to remove the light source and detector from the zone of measurement, thus the interference of temperature and the magnetic field is eliminated and the electric noise is reduced.

A special magnetic conductor traps ferromagnetic debris only in the zones of measurement. After the electromagnet is switched off the magnetic conductor has practically no residual magnetization, so the oil stream becomes able to remove the accumulated particles from the channel after measurements.

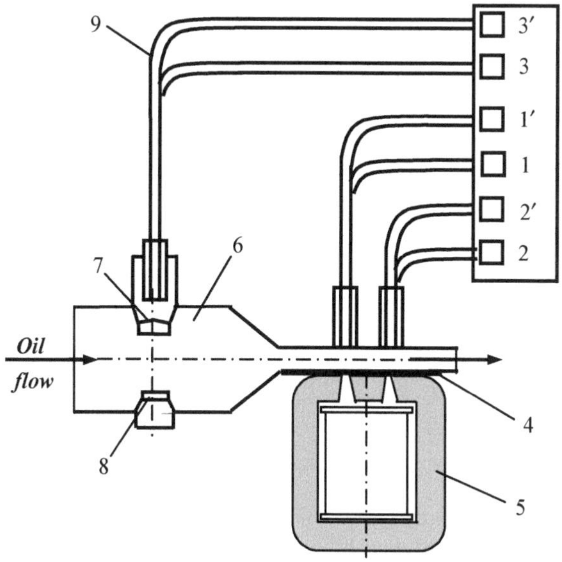

Fig. 5.38 Optical ferroanalyzer sensor: *1, 2, 3* sources of emission; *1′, 2′, 3′* photodetectors; *4* substrate; *5* electromagnet; *6* channel for measuring oil contamination; *7, 8* optical system; *9* optical fibers

The magnitude of the electric signal can be determined from the formula:

$$U = \Phi S, \tag{5.5}$$

where $\Phi = \Phi_0\tau$ is the optical flux incident on the sensitive area of the photodetector; Φ_0 is optical flux reflected by the substrate; τ is transmittance of the optical fiber and S is integrated sensitivity of the photodetector. The value Φ_0 can be determined from the formula:

$$\Phi_0 = EA_{sub}R, \tag{5.6}$$

where E and A_{sub} are the illumination and the area of the measured zone of the substrate, R is the reflectance.

Relations (5.5) and (5.6) allow one to obtain a relationship between the relative changes in voltage at the output of the photodetector and the area of precipitate:

$$\frac{\Delta U}{U} = \frac{U - U_d}{U} = 1 - \frac{U_d}{U} = 1 - \frac{\Phi'_0 \tau S}{\Phi_0} = 1 - \frac{\Phi'_0}{\Phi_0} = 1 - \frac{EA'_{sub}R}{EA_{sub}R} = 1 - \frac{A'_{sub}}{A_{sub}} = \frac{\Delta A_{sub}}{A_{sub}},$$

where U, A_{sub}, Φ_0, U_d, A'_{sub} and Φ_0' are the characteristics of optical radiation when particles have not yet deposited on the substrate and after that.

Thus, the relative voltage changes are equal to the relative change of the surface area occupied by precipitated ferrous particles. It is evident that the value $(\Delta U/U)_1 + (\Delta U/U)_2$ (subscripts 1, 2 relate to the first and second zones) can indicate the concentration of particles in the lubricating material, while the value $(\Delta U/U)_1$—$(\Delta U/U)_2$ characterizes the shift of the size distribution of particles towards the region of larger size.

The oil lubricity is rated by readings of the optical systems before applying the magnetic field to the oil channel and determining the optical density of the oil in each zone. The total contamination of oil (changes in the optical density resulting from oxidation, aging and mechanical impurities) is determined from the mean readings of the optical density of the used oil in respect to the fresh oil.

When oil has low optical density and its contamination changes little in operation (e.g. hydraulic systems), an additional channel *6* is used with a third optical measuring system serving to rate the total oil contamination. This system comprises optical units *7* to collimate the optical radiation at the output of the system *8* and to focus it at the tip of optical fiber *9*. The optical oil density is determined from the formula $D = \lg(U_0/U_1)$, where U_0 and U_1 are the output voltages from the third measuring system during tests of fresh and used oil.

Total measurement system with OF

OF can be applied as a portable device in service centers, in plant laboratories as well as device installed in oil circulation system, where it is necessary to eliminate the bubbling, which distorts the optical signal. It is solved in the OF by vacuum treatment of the oil to be tested. Figure 5.39 shows photo and scheme of the measuring system.

The vacuum pump P delivers the oil periodically with the time intervals Δt_0 through the open valves V1 and V2. The optical measuring system MS-3 serves also for gauging the oil level when it is filled up. After oil is filled up, the valves

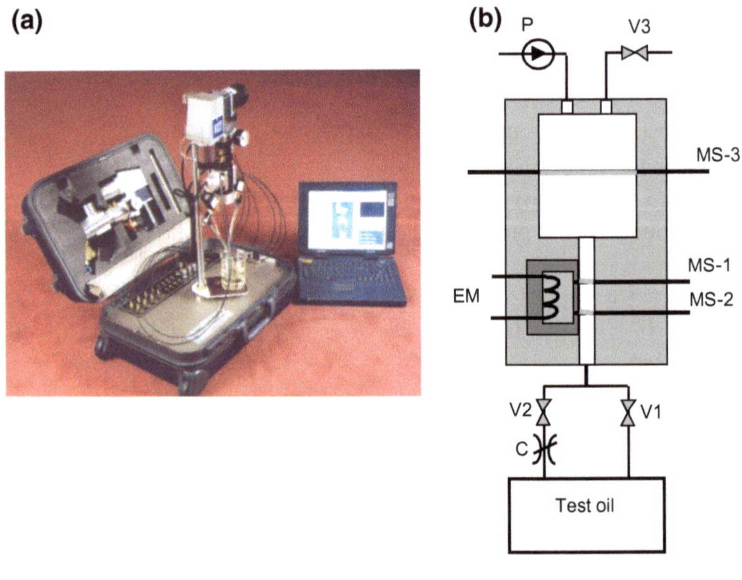

Fig. 5.39 Photo (a) and scheme (b) of optical ferroanalyzer: *P* pump; *EM* electromagnet; *V1*, *V2*, *V3* electromagnetic valves; *C* flow speed controller; *MS-1*, *MS-2*, *MS-3* optical measuring systems

close, and the oil is treated by vacuum during time Δt_1 (3–5 min) determined by its viscosity. After that the pump is switched off and the *j*th measurement is made. Signals $U_{ji}(t)$ from the outputs of measuring systems are counted and processed using the following method. Changes in the optical density of the used oil are determined in each *i*th measuring system against that of pure oil using the formula: $\Delta D_{ji} = \lg \frac{U_{1i}}{U_{ji}(0)}$, where U_{1i} is a value of the output signal of the first measurement (j = 1) in the *i*-th measuring system after a regular oil replacement, i.e. that of pure oil. The mean value of change in the oil density is determined $\Delta D_j = \frac{1}{n}\sum_{i=1}^{n} \Delta D_{ji}$ (*n*—the number of measuring systems; if total contamination is measured by systems 1 and 2, then n = 2, if by system 3, then n = 1) and the results are compared with the allowable contamination level ΔD^*. If $\Delta D_j > \Delta D^*$, then a warning signal is generated and oil should be replaced.

After the electromagnet EM is switched on MS, valves V2 and V3 are opened and the oil passes through the accumulation channel. Oil flow rate (0.5–2 ml/min) is controlled to provide the efficient accumulation depending on the oil viscosity. The debris precipitate in the measuring zones so that larger particles are accumulated in the first zone (i = 1) and finer particles accumulate in the second zone (i = 2). In the course of their accumulation the indications $U_{ji}(t)$ of measuring systems MS-1 and MS-2 were read at $t = \Delta t_2$. The duration of accumulation Δt_2 is determined by the least concentration of the debris in the oil to be registered and by the oil viscosity. The signals $U_{ji}(t)$ diminish as accumulation of the particles proceeds (time of accumulation t becomes longer) in the measuring zones so that the area occupied by the precipitated particles expands, i.e. the area of reflecting portions decreases. After that the signals $U_{ji}(t)$ can serve to calculate the wear indices using the known methods [92, 98]. For example, the indicator of concentration of particles with the size corresponding to the *i*th measuring zone N_{ji}, index of the wear rate W_{rj} and index of the wear intensity W_{sj} can be calculated using the formulas:

$$N_{ji} = \lg[U_{ji}(0)/U_{ji}(\Delta t_2)]; \quad W_{rj} = N_{j1} + N_{j2}; \quad W_{sj} = N_{j1} - N_{j2}.$$

The values of the indicators and indices of each *j*th measurement allow us to judge about the condition of wear of a friction unit at a given moment of time corresponding to the *j*th measurement. The indices of wear from each *j*th measurement are compared with the critical allowable values W_r^*, W_s^*, when they are exceeded a signal is generated commanding to stop the unit in order to prevent severe wear. After each *j*th measurement the electromagnet is switched off, the accumulated wear particles are flushed from the channel with the oil stream. After that the entire process of measurements is repeated.

The operation of OF is fully computerized: control of valves, pump, electromagnet, registration and processing of signals are performed by a processor.

Laboratory Tests of OF

The analyzer was tested following a specially developed technique. For this purpose SAE 10 W/30 oil and the standard magnetic contaminant were used (carbonyl iron powder with size of particles 0.5–12 μm and mean size 5 μm). Oil samples were calibrated by the powders with $C = 5 \times 10^{-4}$ mass%, 10×10^{-4} mass%, 20×10^{-4} mass%, and 50×10^{-4} mass%.

The measuring channel MS-3 was used to measure the total oil contamination. Figures 5.40 and 5.41 show the changes of signals at the outputs of the three measuring systems. The oil flowed through the accumulation channel at a rate 1 ml/min, the accumulation of particles lasted 1 min.

Figure 5.42 shows the indices of the total oil contamination, wear rate and intensity calculated using the above formulas as a function of concentration of ferromagnetic particles in the oil.

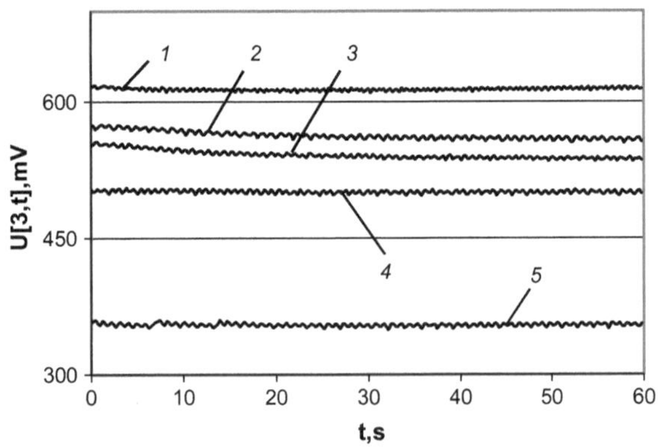

Fig. 5.40 Relationship between the output signal of channel of total oil contamination and concentration of contaminant: *1* pure oil; *2* $C = 5 \times 10^{-4}$ mass%; *3* 10×10^{-4}; *4* 20×10^{-4}; *5* 50×10^{-4} mass %

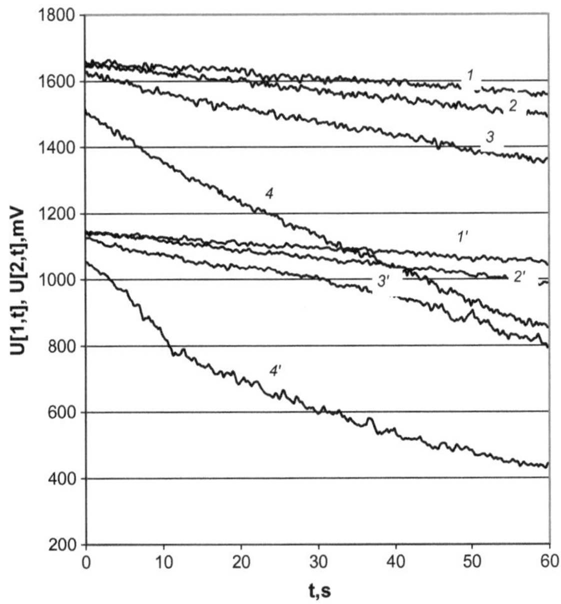

Fig. 5.41 Output signal in first $U[1, t]$ and second $U[2, t]$ accumulation channels as function of concentrations of contaminant: *1*, *1′* 5×10^{-4} mass%; *2*, *2′* 10×10^{-4}; *3*, *3′* 20×10^{-4}; *4*, *4′* 50×10^{-4} mass%

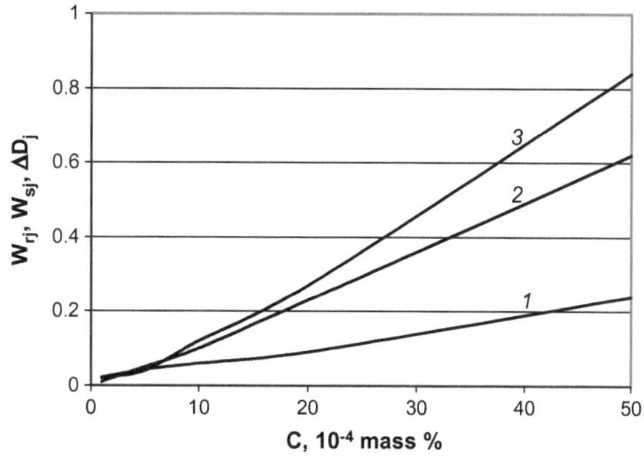

Fig. 5.42 Changes of index of total oil contamination ΔD_j (*1*), index of wear rate W_{rj} (*2*) and index of wear intensity W_{sj} (*3*) as function of changes in concentration of ferromagnetic particles in oil

The above results show that the new OF is highly sensitive to low concentrations (5×10^{-4} mass%) and it allows one to estimate the index of total oil contamination, and the indices of wear rate and intensity.

Application of OF to condition monitoring of tribosystems

The analyzer was used to monitor oils after operation in the internal combustion engine of a heavy transport vehicle. Under high loads the total oil contamination increases quicker as the haul run extends, hence the method of measurement was applied to measuring the total contamination in the accumulation channels. The oil OE 10 W/30 (Shang Young Co., ν = 70 cSt at 40 °C) was used and sampled after 300 km run.

A processor controlled the routine of measurements and processed the results. The oil flow in the accumulation channel was set at 1 ml/min.

Oil analysis from the internal combustion engine. The signals from the outputs of the accumulation channels (i = 1 and i = 2) were read before the electromagnet was switched on (t = 0 min), $U_{j1}(0)$ = 11.7 mV and $U_{j2}(0)$ = 11.8 mV. Then the changes in the optical density of the used oil were contrasted to the pure oil in the measuring channels:

$$\Delta D_{j1} = \lg \frac{U_{11}}{U_{j1}(0)} = 1.276, \ \Delta D_{j2} = \lg \frac{U_{12}}{U_{j2}(0)} = 1.268,$$

where U_{11} = 221 mV and U_{12} = 219 mV are values of output signal of the first measurement (j = 1) after a scheduled oil replacement, i.e. for the pure oil. The mean change in the optical density $\Delta D_j = \frac{1}{2}(\Delta D_{j1} + \Delta D_{j2}) = 1.272$ was compared with the allowable contamination of the oil in the engine ΔD^* = 1.5. In this jth measurement it was $\Delta D_j < \Delta D^*$, still the values were quite identical proving that the oil was approaching the critical condition, i.e. that it was about the time to replace it. After that the electromagnet was switched on to start the process of accumulation, the readings $U_{j1}(t)$ and $U_{j2}(t)$ of the measuring systems were registered during time Δt_2 = 10 min (Fig. 5.43a). After that the electromagnet was

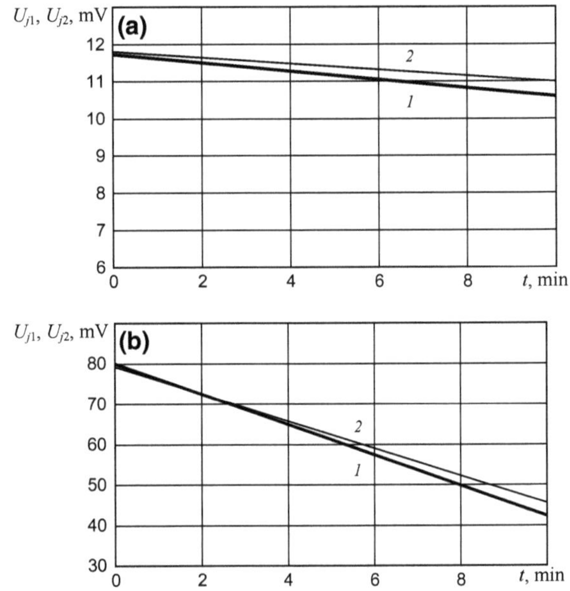

Fig. 5.43 Relationship of output signals from first $U[1, t]$ (*1*) and second $U[2, t]$ (*2*) accumulation channels when analyzing oil from internal combustion engine (**a**) and transmission gear (**b**) after 300 km run of heavy duty transport vehicle

switched off and the accumulated particles were flushed away from the channel by the oil stream.

The indicators of concentration of the particles with the sizes corresponding to the input N_{j1} and output N_{j2} accumulation channels, wear rate index W_{rj} and wear intensity index W_s were determined:

$$N_{j1} = \lg[U_{j1}(0)/U_{j1}(\Delta t_2)] = 1.10; \quad W_{rj} = \tfrac{1}{2}\left(N_{j1} + N_{j2}\right) = 1.09;$$
$$N_{j2} = \lg[U_{j2}(0)/U_{j2}(\Delta t_2)] = 1.07; \quad W_{sj} = \left(N_{j1} - N_{j2}\right) = 0.03.$$

After that the wear indices were compared with the limits allowable for the engine $W_r^* = 2$ and $W_s^* = 1$. It is apparent that Wrj and Wsj are quite below the allowable parameters, i.e. the process of wear proceeds in a normal steady mode.

Analysis of the oil from the transmission gear. Messasurements were performed similarly like with the oil from the engine. The values of the output signal when the system is filled up with fresh oil in the channels amounted to $U_{11} = 235$ mV and $U_{12} = 225$ mV. When the system was filled with the oil used during 300 km run of the transport vehicle the sig-nals from the outputs of the measuring accumulation zones before actuation of the electromagnet ($t = 0$) amounted to $U_{j1}(0) = 82$ mV and $U_{j2}(0) = 81$ mV. The change in the optical density of the used oil in respect to the fresh oil in the accumulation channels was

$$\Delta D_{j1} = \lg\frac{U_{11}}{U_{j1}(0)} = 0.457, \ \Delta D_{j2} = \lg\frac{U_{12}}{U_{j2}(0)} = 0.444.$$

The mean change in the optical density was

$$\Delta D_j = \frac{1}{2}(\Delta D_{j1} + \Delta D_{j2}) = 0.45.$$

The allowable contamination of the oil in the transmission gear is $\Delta D^* = 1.3$, but ΔD_j in this measurement was much less than the allowable value, i.e. the condition of the oil was satisfactory. Figure 5.43b presents the readings $U_{ji}(t)$ by the measuring systems $U_{j1}(t)$ and $U_{j2}(t)$ after operation of the electromagnet during the time $\Delta t_2 = 10$ min. In this case the indicators of the concentration of particles with the dimensions corresponding to the first N_{j1} and the second N_{j2} measuring zones, wear rate index W_{rj} and wear intensity indicator W_{sj} are equal to:

$$N_{j1} = lg\,(U_{j1}(0)/U_{j1}(\Delta t_2)) = 1.84 \quad W_{rj} = 1/2\left(N_{j1} + N_{j2}\right) = 1.78$$
$$N_{j2} = lg\,(U_{j2}(0)/U_{j2}(\Delta t_2)) = 1.72 \quad W_{sj} = \left(N_{j1} - N_{j2}\right) = 0.12.$$

The ultimate allowable indicators of wear of the transmission gear are $W_r^* = 4$ and $W_s^* = 1$. It is apparent that W_{rj} and W_{sj}. are below the allowable values, i.e. the process evolves in a normal steady mode.

The analysis of the presented case shows that oil becomes contaminated more intensively in the engine due to more severe heating than in the transmission gear, but the friction surfaces in the gear wear more intensively due to heavier contact loading. Yet, regular oil replacements and maintenance of this transport vehicle are scheduled after each 1200 km run, disregarding the features of operation of friction units. The results of the analysis of the oil in the engine show that it loses its performance before replacement and may upset normal operation of the engine.

Thus, optical ferroanalyzer allows one to monitor the conditions of tribosystems by analysing the debris in lubricant and to rate the total oil contamination. The results of the OF tests have shown that it is sensitive to very low concentrations of ferromagnetic particles, so it can serve as an instrument for real-time wear monitoring of tribosystems.

5.5.3 *Inductive Wear Particle Counter*

Opto-magnetic detector and optical ferroanalyzer presented earlier (paragraphs 5.5.1 and 5.5.2) can estimate the mass concentration of the small particles of wear. But in real practice in heavy-loaded tribosystems it is very important to monitor the appearance of the large particles (>100 μm) which evidence the start of severe wear and at the same time can accelerate wear due to abrasive effect. For example at severe wear of the Babbit linings of the sliding bearings the content of non-ferrous debris is quickly growing. Lining wear can result in the accelerated wear of the shaft and simultaneous appearance of ferrous debris. Reliable detection of large non-ferrous particles in such important tribosystems as heavy compressors and airplane engines is a hot problem in predictive maintenance.

Large ferrous particles are detected by the magnetic plugs in the lubrication system. For detection of non-ferrous ones the sensor can be used detecting the appearance of the electrical contact in the gap of metal plates by its bridging with metal particles [99]. It can generate the signal also in case of water bridging the gap, but it cannot calculate the number of particles.

Size and concentration of metal particles in liquid can be detected by the inductive sensor.

The principle of the inductive sensor

Wear particles in the flow of oil change the magnetic field depending on the conductivity of particle, and its magnetic permeability. The ferrous particles have positive relative magnetic permeability and it increases the external field intensity. Non-ferrous metal particles have relatively low magnetic permeability and their magnetic effect is determined by conductivity, and resulting eddy currents decrease the external magnetic field.

The measuring unit includes three coils—sensitive one (SC), and the coils C1 and C2 generating the alternative magnetic field in the coil SC (Fig. 5.44a). If there are no particles in the oil the effects of the coils C1 and C2 are compensated. When

Fig. 5.44 The measuring coil unit (**a**) and signal generated in sensitive coil by ferrous (**b**) and non-ferrous (**c**) particles. *C1* and *C2* exiting coils, *SC* sensitive coil

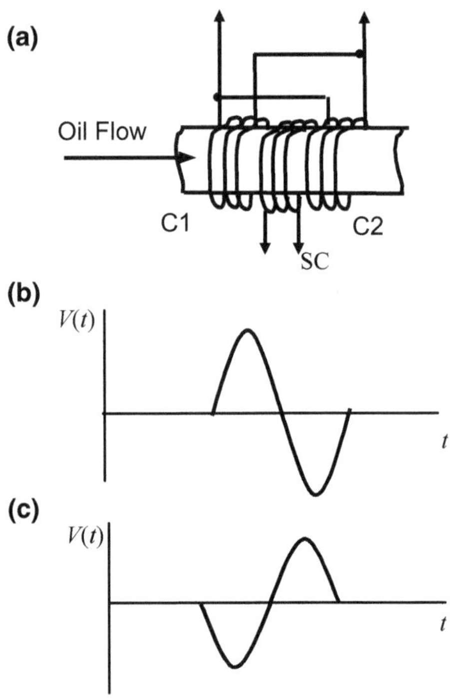

the ferrous particle passes there is an increase in the field exited by the coil C1 in the coil SC and the pulse of positive polarity is registered (Fig. 5.44b). In case of non-ferrous particle such a pulse is negative due to the reaction of the coil C2 (Fig. 5.44c). The phase of the resulting signal provides the information on the particle material and the signal amplitude relates to the particle size.

The Figs. 5.45 and 5.46 illustrate the signal generation and processing [75]. The shape of the signal $V_1(t)$ sent to the coils C1 and C2 from the generator is shown in Fig. 5.45a. In case of ferrous particle the sensing coil SC generates the signal of the same frequency modulated by low-frequency component (Fig. 5.45b) determined by the particle passage through the coils. The signal amplified in the preliminary amplifier *3* (Fig. 5.46) comes to the multiplier *4* and mixed with the generator *1* signal. As a result the pulse high-frequency signal is formed with the same phases of the low-frequency and high-frequency components (Fig. 5.45c). This signal comes to the cascade *5* blocking the constant component of the signal, and then to the low–pass filter *6*. The output signal of the filter contains the information on the particle (Fig. 5.45d). It comes to the final amplifier *9* and to the discriminator *10* which distributes the signals of ferrous and non-ferrous particles to the counting channels *11* and *12*.

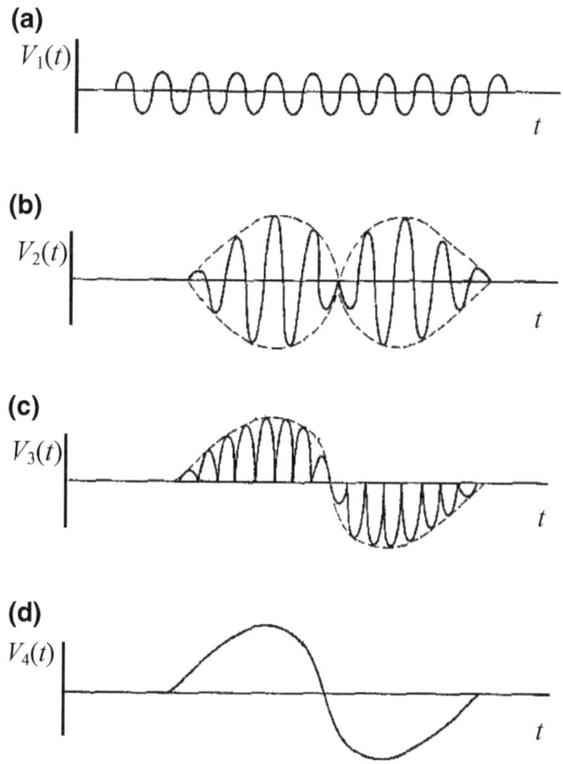

Fig. 5.45 Signals: **a** output from the driving generator output $V_1(t)$; **b** from sensitive coil $V_2(t)$, **c** from multiplier $V_3(t)$; **d** from low-pass filter $V_4(t)$

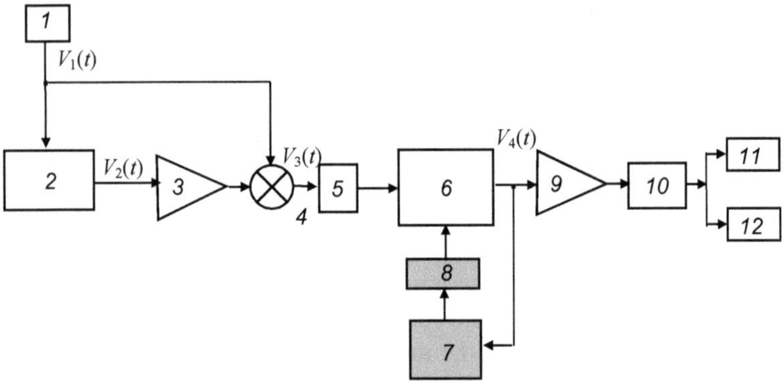

Fig. 5.46 Block scheme of inductive sensor: *1* driving generator, *2* measuring coil unit, *3* preamplifier, *4* multiplier, *5* the cascade blocking the constant component of the signal, *6* low–pass filter, *7* detector of the pulse duration of the particle coming through the coils, *8* the controller of the low-pass filter, *9* final amplifier, *10* discriminator, *11* ferrous particle counter, *12* non-ferrous particle counter

In order to improve the sensitivity of the device there was found useful to increase the ratio of signal to noise. The low-pass filter (Fig. 5.46) is can be adjusted to the width of the frequency band using the detector *7* of the pulse duration of the particle coming through the coils and the controller *8* of the low—pass filter.

Coefficient of the transfer of the controller *8* can be written as $k(\tau_p) = F\tau_p$, where *F*—proportionality coefficient; τ_p—pulse duration, $\tau_p = L_{coil}/v_p$, L_{coil}—length of coil channel, v_p—velocity of particle.

Transmittance band width Δf of the filter *6* is fixed relating to the average velocity of the oil in the flow ($\Delta f \sim v_{oil}/L_{coil}$, where v_{oil}—average oil velocity). Duration of the pulse of misbalance signal is determined by the sizes of coils C1, C2, and SC, as well as velocity of particles in the oil flow through the coils. In case of small particles their velocity is equal to flow velocity. Controller of the transmittance band *8* determines the cut-off of the low-frequency filter *6* in accordance with the pulse signal duration. As a result the optimum tuning of the filter is carried out, signal-to-noise ratio and device sensitivity are increased. Figure 5.47a presents the signal at the output of low-pass filter *6* without the frequency band optimization and it is clear that signals of small amplitude cannot be extracted from the total signal. Figure 5.47b shows the same signal after filter optimization and evidences registration of the signals of small amplitude relating to small particles of wear. The photo of the inductive sensor is presented in Fig. 5.48.

Laboratory tests of the inductive sensor

The tests have been carried out at monitoring of the piston compressor at the Belarus gas-processing plant. The crankshaft of the compressor with the Babbitt bearing linings was a target of monitoring in order to detect the start of the linings

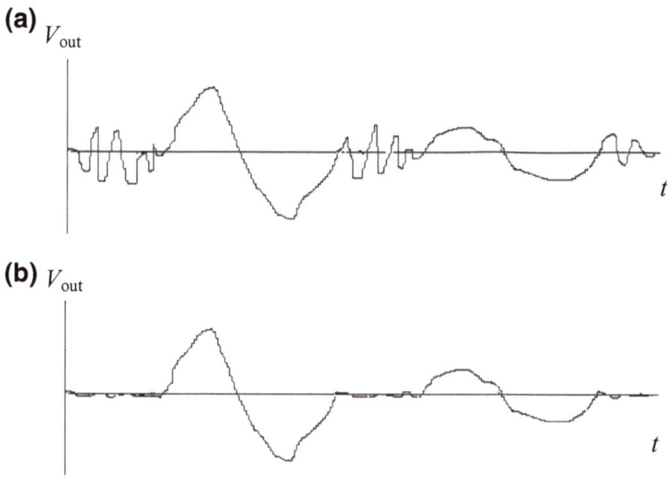

Fig. 5.47 Inductive sensor output without optimization (**a**) and with optimization (**b**) of the frequency band of low—pass filter

Fig. 5.48 Photo of the inductive sensor

severe wear and necessity of their replacement. The routine technique of monitoring the compressor was the use of the electric detector of the metal chips [64]. The inductive sensor and the optical microscopy were used as additional methods for debris analysis in the oil.

During the tests the urgent stop of compressor was done due to the signal of the electric chip sensor. The machine was stopped and it was found that the bearings were damaged severely.

The debris sediment in the electric sensor of wear installed in the compressor was studied. The sediment was washed by hexane and deposited on the glass plate. It was shown that both large (Fig. 5.49a) and small particles (Fig. 5.49b) are present.

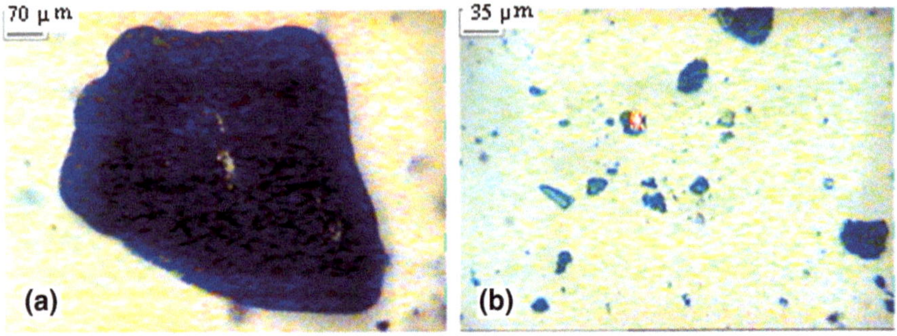

Fig. 5.49 Micro photos of the particles in the deposit on the plates of the electric sensor

While studying the oil sample taken 72 h before failure with the use of inductive sensor in laboratory the Babbitt particles with size 200–1000 μm were registered. Typical record at the output of the sensor with the pulses relating to 500 and 300 μm is presented in Fig. 5.50.

The oil sample taken 72 h before failure was also studied by optical microscope MBR-1 combined with the image analyzing computer. It was shown that the sample contains many small (<50 μm) transparent (quartz and sand) particles and ferrous ones (Fig. 5.51a). There were found also large (up to 800 μm) ferrous and non-ferrous particles (Fig. 5.51b, c).

Later on after changing the bearings large particles are absent in the oil samples and only the small ones (<50 μm) are present (Fig. 5.52a, b).

Henceforth, it was found that the inductive sensor can efficiently detect the initial stage of the severe wear of the bearing linings, and thus providing the predictive maintenance of machinery.

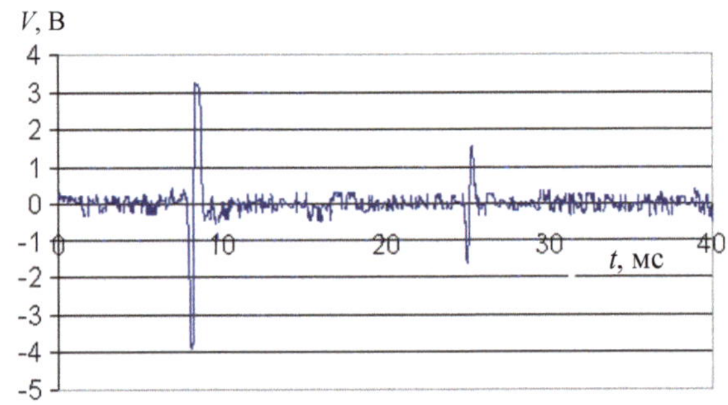

Fig. 5.50 Typical record of the output signal of the inductive sensor (oil sample taken 72 h before failure)

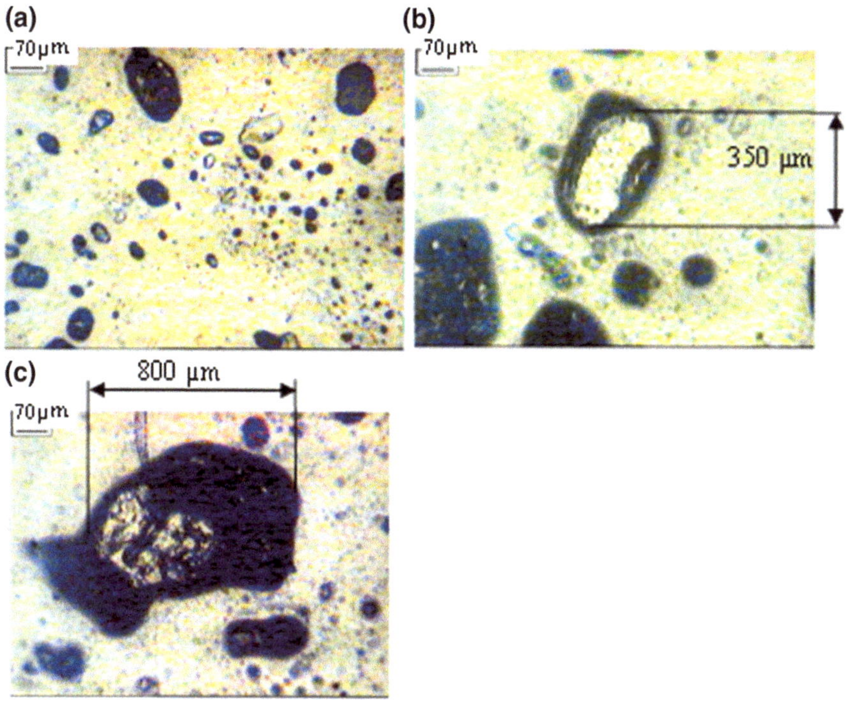

Fig. 5.51 Micro photos of small (**a**) and large (**b**, **c**) wear particles in oil sample 72 h before failure of the bearings

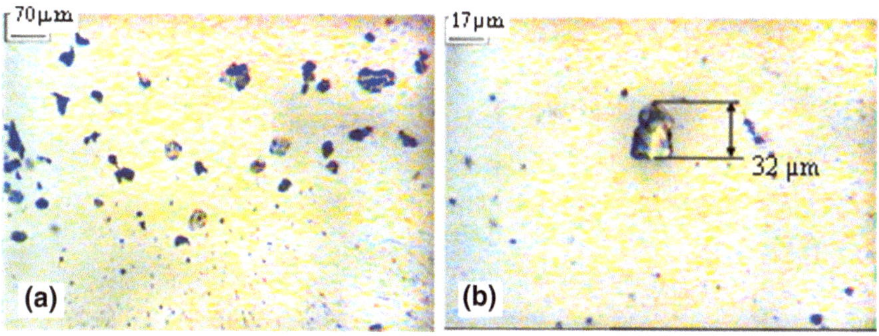

Fig. 5.52 Micro photos of the particles in the samples taken after 110 days of operation (**a**) and 140 days of operation (**b**) after changing the bearings

5.6 Conclusions

It is possible to evaluate the condition of machines and predict their wear on the basis of wear particles concentration and size distribution. Observation of the wear particles yields also information on the type of wear and often on the specific location of the worn part. If it is necessary to have maximal information on the wear particles it is advisable to use methods such as ferrography and RDS. These methods can also be used in the large maintenance centers. The complex built-in systems are justified for use in expensive and critical equipment (e.g. airplanes and helicopters).

When complete information on the wear particles is not required, the simpler and less expensive field diagnosis methods are suitable. However field diagnostics should be based on an appropriate data base created with the use of more complex diagnosis methods. This data base for each specific machine type, oil type, and operation conditions makes it possible to use the relatively simple field methods based on the determination of certain characteristic of the oil (for example, the optical density, magnetic permeability, and so on). At present time the most promising field methods are the magneto-optical ones, which use the instruments combining simplicity and portability with adequate accuracy of measurement. In the selection of a particular method it is necessary to consider the importance of the machine, its cost, operation conditions (loading regime, degree of contamination of the oil), and the diagnostic objective.

The efficiency of diagnosis depends greatly on the type and size of wear debris. Mechanisms subjected to monitoring generate wear particles of different types and sizes, and one technique can not provide efficiency in all cases, so several methods should be used.

Efficient monitoring methods and devices were developed in Metal-Polymer Research Institute of National Academy of Sciences of Belarus. In some cases these methods were developed and tested in cooperation with the Korea Institute of Science and Technology.

References

1. V.C. Westcott, *Monitoring of Wear in Fundamentals of Tribology* (MIT Press, Cambridge MA, 1978), pp. 811–829
2. J.H. Johnson, in *Monitoring of machine wear by used oil analysis. Proceedings of International Conference on Fundamentals of Tribology* (Cambridge, Massachusetts, 1978)
3. F.F. Lockwood, Dalley, *Lubricant Analysis, 18th edition, Metals Handbook, Tribology*, vol 19 (ASM, 1992), pp. 300–312
4. D.P. Anderson, *Wear particle atlas (revised). Naval Air Eng. Center, Report NAEC-92-163* (Foxboro Edition, Burlington, 1982), p. 192
5. D.C. Schalcosky, C.S. Byington, Advances in real time oil analysis. Practicing Oil Analysis 11 (2000). Available: http://www.machinerylubrication.com/Read/138/real-time-oil-analysis

6. R. Bowen, D. Scott, W. Seifert, V. Westcott, Ferrography. Trib. Int. **9**(3), 109–115 (1976)
7. G. Pocock, The observation of spherical debris from a failed soft metal bearing. Wear **38**(1), 189–191 (1976)
8. J. Kleis, U. Muiste, U. Pilvre, H. Uuemois, H. Uetz, The physical mechanism of the formation of metal microspheres in the wear process. Wear **53**(1), 79–85 (1979)
9. K. Nakayama, J. Okamoto, Wear particle distribution and its relation to the wear transition under boundary lubrication. Wear **70**(1), 125–129 (1981)
10. J. Ding, Determining fatigue wear using wear particle analysis tools. Practicing Oil Analysis 9 (2003). Available: http://www.machinerylubrication.com/Read/526/fatigue-wear-particle-analysis
11. T. Kayaba, K. Kato, T. Akagaki, Ferrographic study of wear (second report) – the valuation of wear conditions with a magnetic wear separator. J. JSLE Int. Ed. **1**, 53–58 (1986)
12. A.A. Reda, R. Bowen, V.C. Westcott, Characteristics of particles generated at the interface between sliding steel surfaces. Wear **34**(3), 261–273 (1975)
13. P.B. Senholzi, Oil analysis/wear particle analysis. Tribology 1978: Materials Performance and Conservation (Swansea, 1978), pp. 129–136
14. B.J. Roylance, G. Pocock, Wear study through particle size distribution I: application of the Weibull distribution to ferrography. Wear **90**(1), 113–136 (1983)
15. P.T. Barwell, The contribution of particle analysis to the study of wear of metals. Wear **90** (1), 167–181 (1983)
16. O.P. Sondhiya, A.K. Gupta, Wear debris analysis of automotive engine lubricating oil using by ferrography. Int. J. Innov. Res. Sci. Eng. Technol. **2**(5), 46–54 (2012)
17. D. Scott, V.C. Westcott, Predictive maintenance by ferrography. Wear **44**(1), 173–182 (1977)
18. R.T. Lewis, Application of magnetization measurements to wear debris analysis. Wear **74**, 177–190 (1981)
19. ASTM D7684–11: Standard Guide for Microscopic Characterization of Particles from In-Service Lubricants
20. Z. Peng, T.B. Kirk, Wear particle classification in a fuzzy grey system. Wear **225–229**(Part 2), 1238–1247 (1999)
21. N.K. Myshkin, A.Ya. Grigoriev, L.V. Markova, Condition monitoring of tribosystems by wear debris analysis. Int. J. of Applied Mechanics and Engineering **7**(3), 923–947 (2002)
22. G.W. Stachowiak, P. Podsiadlo, Surface characterization of wear particle. Wear **225–229** (Part 2), 1171–1185 (1999)
23. T.B. Kirk, D. Panzera D, R.V. Anamalay, Z.L. Xu, Computer image analysis of wear debris for machine condition monitoring and fault diagnosis. Wear **181–183**(Part 2), 717–722 (1995)
24. H. Wu, N.M. Kwok, S. Liu, T. Wu, Z. Peng, A prototype of on-line extraction and three-dimensional characterisation of wear particle features from video sequence. Wear **368–369**, 314–325 (2016)
25. T. Wu, Y. Peng, S. Wang, F. Chen, N. Kwok, Z. Peng, Morphological feature extraction based on Multiview images for wear debris analysis in on-line fluid monitoring, Tribol. Trans. Jun 2016, pp. 1–11. http://dx.doi.org/10.1080/10402004.2016.1174325
26. L.V. Markova, N.K. Myshkin, Tribojunction diagnostics based on wear particles. Trenie Iznos **9**(6), 1109–1118 (1988)
27. N.K. Myshkin, O.V. Kholodilov, L.V. Markova, M.S. Semenyuk, Diagnostics of lubricated moving junction wear. Trenie Iznos **7**(6), 1091–1101 (1986)
28. T. Hansen, P. Skiba, C.R. Hodkins, Determination of iron in used lubricating oils by spectrochemical analysis. Anal. Chem. **23**(10), 1362–1365 (1951)
29. L.R.P. Butler, D.B.O. Villiers, Spectrometric oil analysis for tracing unusual wear in lubricated systems. S. Afr. Mech. Eng. **23**(12), 261–266 (1973)
30. E.J. McCormack, N. Oil analysis, Prod. Eng. **28**(4), 82–85 (1981)
31. L.L. McCorriston, W.M. Meston, Improved wear metals analysis for lubricant field trials. SAE Techn. Pap. Ser. 1979. N 790935

32. T.M. Hunt, *Handbook of wear debris analysis and particle detection in liquids* (Elsevier Science, London and New York, 1993), p. 488
33. M. Lukas, R.J. Yurko, Current technology in oil analysis spectrometers and what we may expect in the future. Integrated monitoring, diagnostics and failure prevention. in *Proceedings of a Joint Conference*, Mobile, Alabama, pp. 161–171, 1996
34. H. Levinsohn, Limitations of atomic absorption spectrophotometry applied to spectrometric oil analysis. ASLE Trans. **27**(1), 24–32 (1984)
35. G. Cumming, I.G. McDonald, The determination of iron in lubricatisng oil by X-ray fluorescence spectrometry. Wear **103**(1), 57–66 (1985)
36. J.L. Berg, J.E. Lieser, Lubricant testing as an oil to bearing damage analysis. Lubr. Eng. **29** (12), 539–543 (1973)
37. Z. Yang, X. Hou, B.T. Jones, Determination of wear metals in engine oil by mild acid digestion and energy dispersive X-ray fluorescence spectrometry using solid phase extraction disks. Talanta **59**(4), 673–680 (2003)
38. G.R. Humphrey, Condition-based monitoring using energy dispersive X-ray fluorescence. Lubr. Eng. **55**(10), 19–26 (1999)
39. D. Severin, O. Glinzer, E. Wedepdil, U. Hildebrandt, Zur Mesung von motorenolverschmutzung. Schmierungstechnik + Tribologie **28**(3), 87–88 (1981)
40. B. Carbonel, Lubrification des roulements detection et analyse de la pollution par particles. Mec. Mater. Elec. **59**(316), 40–41 (1976)
41. S. Norose, M. Sato, T. Sasada, Ferrographic analysis of wear particles produced under lubrication condition (Part I). J. JSLE Int. Ed. **5**, 35–40 (1984)
42. D. Scott, V.C. Westcott, Recent developments in machinery condition monitoring by ferrography. in *Proceedings of the Cconference "Patterns of Tribology"*, Paisley, pp. 503–509, 1979
43. B.J. Roylance, Ferrography–then and now. Trib. Int. **38**(10), 857–862 (2005)
44. A. Lobarzewska, Zastosowanie metody fotoelektryczney do oceny wiasnosci dysperguja-cych olejow smarowych. Techn. Smarown. **4**, 117–121 (1973). (in Poland)
45. M.C. Isa, N.H.N. Yusoff, H. Naina, M.S.D. Yati, M.M. Muhammada, I.M. Nor, Ferrographic analysis of wear particles of various machinery systems of a commercial marine ship. Procedia Eng. **68**, 345–351 (2013)
46. L.A. Shabalinskaya, V.V. Golovanov, E.S. Bubnova, L.V. Milinis, The Study of the possibility of the tribodiagnostics of the technical condition of the helicopter main reduction gear. J. Friction Wear **36**(1), 96–101 (2015)
47. A.L. Price, B.J. Roylance, The rotary particle depositor a response to problems experienced with wear particle deposition. Proceedings of International Conference on Condition Monitoring (Swansea, 1984) pp. 596–607
48. A.L. Price, B.J. Roylance, L.X. Zi, The PQ—a method for the rapid quantification of wear debris. in *Proceedings of International Conference on Condition Monitoring* (Swansea, 1987) pp. 391–405
49. N. Myshkin, Studies of wear debris in oils by magnetic method. Colloid J. **42**(5), 988–991 (1980)
50. M. Lukas, D.P. Anderson, T. Sebok, D. Filicky, LaserNet Fines—a new tool for the oil analysis toolbox. Practicing Oil Analysis 9 (2002). Available: http://www.machinerylubrication.com/Read/383/lasernet-fines-oil-analysis
51. Device for detecting and monitoring debris in oil (on-line ferrograph), Pat. JP 4-49902 G01 N 15/02. Published 12 Aug 1992
52. F. Colladon, Pat. US 9,248,454 B08B 5/00, B03C 1/28, B03C 1/032. Device and method for recovering magnetic particles trapped on a magnetic plug—No. 20120204910. Filed 16.08.12. Published 02 Feb 2016
53. R. Kudrna, Pat. Application US20150177116, G01 N 15/10, G01N015/10, H01F 7/02, G01 N 27/74. Magnetic chip detector/collector. Published 25 Jun 2015

54. J-P.Y. Bares, E.Y. Blanchard, US 5027065 F16 N 29/00, G01 N 027/74, G01R 033/12. Particle sensor plug with wireless casing connection—No. 07/524,491. Filed 17.05.90. Published 25 Jun 1991
55. W.E. Rumberger, Pat. US5118410 B01D 35/14, G01 N 15/06, F01 M 11/03. Debris collecting and measuring device with induction coil screen—No. 07/544,941. Filed 28.06.90. Published 02 Jun 1992
56. C.E. Freese V, M.J. Rostoskey, R.E. Garvey III, Pat. US 5604441 G01 N 27/22, G01 N 33/26, F01 M 11/03, G01R 027/26. In-situ oil analyzer and methods of using same, particularly for continuous on-board analysis of diesel engine lubrication systems—No. 08/403,638. Filed 14.03.95. Published 18 Feb 1997
57. F.J-C. Makowski, A. Thenaisie, Pat. FR2667153 F16N29/00, G01N21/53, G01V8/16. Sensor for impurities in a fluid and its use in a circuit - No. FR19900011843. Filed 26.09.90. Published 27 Mar 1992
58. R.W. Bogue, An improved magnetic plug for the continuous monitoring of wear debris. in *Proceedings of International Conference on Condition Monitoring* (Swansea, 1984) pp. 628–635
59. M. Montuschi, M. Incardona, Pat. EP0290397 F16N29/00, G01N15/06. Sensor for detecting the ferromagnetic-particle content of a fluid—No. EP19880830071. Priority 06.03.87. Published 09 Nov 1988
60. M.S. Ozogan, A.I. Khalil, P.S. Katsoulakos, Tribological failure detection and condition monitoring for diesel engines. Wear **130**(1), 189–201 (1989)
61. K.W. Chambers, Pat. EP0223400, G01F1/56, G01N15/06, G01N27/72, G01N33/28. Ferromagnetic wear detector and method—No. EP19860307862. Filed 11.10.85. Published 27 May 1987
62. S. Itomi, Pat. US 7,151,383 G01R 27/08, G01 N 33/20. Oil condition sensor—No. US 2005021253, Filed 29.09.05. Published 19 Dec 06
63. T.E. Tauber, P.L. Howard, QDM—a smart on-line debris monitoring system. in *Proceedings of International Conference on Condition Monitoring* (Swansea, 1984) pp. 617—627
64. E. Tsaprazis, Pat. US 4831362 F16 N 29/00, G01 N 33/26, G01 N 15/06, G01 V 13/00, G08B 029/00, G01R 035/00, G01 N 027/74. Failure detection system for a metallic debris detection system—No. 06/863,838. Filed 16.05.86. Published 16 May 1989
65. M. Bradford, Pat. US 4598280 F16 N 29/00, G07C 3/00, G01R 027/02, G01R 033/12, G01 N 015/06. Electric chip detector—No. 06/642,302. Filed 20.08.84. Published. 01 Jul 1986
66. Y. Liu, S.Z. Wen, Y.B. Xie, F. Zhao, Advances in research on a multichannel on-line ferrograph. Trib. Int. **30**(4), 279–282 (1997)
67. T.H. Wu, J.H. Mao, J.T. Wang, J.Y. Wu, Y.B. Xie, A new on-line visual ferrograph. Tribol. Trans. **52**(5), 623–631 (2009)
68. T. Wu, J. Wang, Y. Peng, Y. Zhang, Description of wear debris from on-line ferrograph images by their statistical color. Tribol. Trans. **55**(5), 606–614 (2012)
69. Y. Liu, Z. Liu, Y. Xie, Z. Yao, Research on an on-line wear condition monitoring system for marine diesel engine. Trib. Int. **33**(12), 829–835 (2000)
70. W.E. Rumberger, C.B. GROSS, Pat. EP0572730 B01D35/143, G01N15/06, F01M11/03. Wear detector—No. EP19920304977. Filed 28.06.90. Published 08 Dec 1993
71. D.E Muir, B. Howe, In-line oil debris monitor (ODM) for the advanced tactical fighter engine. in *Proceedings of a Joint Conference "Integrated monitoring, diagnostics and failure prevention"* (Mobile, Alabama, 1996) pp. 111—118
72. H. Zhan, Y. Song, H. Zhao, J. Gu, H. Yang, S. Li, Study of the sensor for on-line lubricating oil debris monitoring. Sens. Transducers **175**(7), 214–219 (2014)
73. W.A. Veronesi, A.P. Weise, R.W. Reed, H.I. Ringermacher, Pat. US4926120 G01 V 3/10, G01 V 003/00, G01R 033/12. In-line metallic debris particle detection system—No. 07/290,531. Filed 2712.88. Published 15 May 1990

74. E. Becker, T. Knoell, R. Hoelzl, Pat. US 8,354,836 G01 N 27/72, G01R 33/12. Device and process for detecting particles in a flowing liquid – No. 20080150518. Filed 26.06.08. Published 15 Jan 2013
75. L.V. Markova, N.K. Myshkin, M.S Semenjuk, V.M Makarenko, A.V. Kolesnikov, V.I. Kokleev, Pat. RU2234080 G01N27/74. Device for detection of metal particles in a flow of lubricating stuff – No. RU20030113971. Filed 12.05.03. Published 10 Aug 2004
76. W. Hong, S. Wang, M. Tomovic, H. Liu, X. Wang, A new debris sensor based on dual excitation sources for online debris monitoring. Meas. Sci. Technol. **26**(9) 095101 (12 pp) (2015)
77. D.W. Singleton, Pat. EP0159094 G01N15/06, G01N27/04, G01N33/28, G01V3/02. Apparatus for detecting metal debris particles—No. EP19850200566. Filed 17.04.84. Published 23 Oct 1985
78. S. Murali, X.G. Xia, A.V. Jagtiani, J.E. Carletta, J. Zhe, Capacitive coulter counting: detection of metal wear particles in lubricant using a microfluidic device. Smart Mater. Struct. (2009). doi:10.1088/0964-1726/18/3/037001
79. J. Zhe, L. Du, J.E. Carlett, R.J. Veillette, Pat. US8,522,604 G01 N 27/74, Metal wear detection apparatus and method employing microfluidic electronic device – No.US 20100109686. Filed 30.10.09. Published 03 Sept. 2013
80. J. Otokar, Pat. EP0254882 G01N29/04. Particle detector – No. EP19870109243. Filed 07.07.86. Published 03.02.88
81. J.G. Miller, R.E. Clark, M.S. Conradi, D.R. Dietz, J.S. Heyman, Pat. US 4,015,464 G01 N 29/032, G01 N 29/02, G01 N 29/036, G01 N 015/00. Ultrasonic continuous wave particle monitor—No. 05/551,913. Filed 21.02.75. Published 05.04.77
82. M. Williamson, The low-down on particle counters. Practicing Oil Analysis 7 (2002). Available: http://www.machinerylubrication.com/Read/351/particle-counters
83. B.T. Kuhnell, J.S. Stecki, Z. Przelozny, X. Huang, C.E. Doran, Pat. US5790246 G01 N 15/02, G01 N 15/06, F01 M 11/03, G01 N 033/28, G01 N 021/00, G01T 001/167. Apparatus and network for determining a parameter of a particle in a fluid employing detector and processor—No. 08/634,207. Filed 18.04.96. Published 04 Aug 1998
84. O.K. Kwon, H.S. Kong, H.G. Han, E.-S. Yoon, N.K. Myshkin, L.V. Markova, M.S. Semeniouk, Pat. US6151108 G01 N 15/02, G01 N 21/03, G01 N 21/59, G01 N 21/05;, G01 N 33/26, G01 N 33/28, G01 N 21/01.On-line measurement of contaminant level in lubricating oil—No. 09/386,168. Filed 31.08.99. Published 21 Nov 2000
85. I. Nelson, Pat. US5982847 G01 N 23/22. Compact X-ray fluorescence spectrometer for real-time wear metal analysis of lubricating oils – No. 08/958,415. Filed 27.10.97. Published 09 Nov 1999
86. B.W. Wilson, S. Price, In-line X-ray fluorescence spectroscopy. Lubr. Fluid Power. 2000, 16–19
87. K.R. Astley, P. Anuzis, I.C.D. Care, T.A. Moore, P.G. Morris, P.D. Rees, Pat. US6794865 G01R 33/563, G01R 33/54, G01 V 003/00. Monitoring the health of a fluid system—No. 09/949,671. Filed 12.09.01. Published 21 Sept 04
88. L.V. Markova, M.S. Semenyuk, Main trends in developing methods and built-in devices for wear diagnosis of tribosystems. J. Friction Wear **17**(3), 82–86 (1996)
89. O.K. Kwon, H. Kong, H-G Han, E-S Yoon, N.K. Myshkin, L.V. Markova, M.S. Semeniouk, Pat. US6151108 G01 N 33/28. On-line measurement of contaminant level in lubricating oil—No. 09/386,168. Filed 31.08.99. Published 21 Nov 2000
90. L.V. Markova., N.K. Myshkin, M.S. Semenyuk, O.K. Kwon, H. Kong, Wear monitoring with on-line optico-madnetic detector. in *Proceedings of the 11th International Conference on Condition Monitoring and Diagnostic Engineering Management*, vol. II, (Laurceston, Austalia, 1998), pp. 585–590
91. N.K. Myshkin, L.V. Markova, M.S. Semenyuk, O.K. Kwon, H. Kong, On-line opto-magnetic detector and its application for air compressor system monitoring. in *Proceedings of the 12th International Colloquium "Tribology 2000-Plus"*, vol. III, (Esslingen, 2000), pp. 1571–1576

92. N.K. Myshkin, L.V. Markova, M.S. Semenyuk, H. Kong, H.-G. Han, E.-S. Yoon, Wear monitoring based on the analysis of lubricant contamination by optical ferroanalyzer. Wear **255**(7–12), 1270–1275 (2003)
93. L.V. Markova, Problems of magneto–optic diagnostics of the wear of lubricated moving junctions. Trenie Iznos **11**(2), 338–342 (1990)
94. A. Einstein, M. Smoluchowski, Brownian motion (ONTI, Moscow, 1936), p. 607 (in Russian)
95. V.E. Fertman, Magnetic fluids (Minsk, Nauka, 1988), p. 184 (in Russian)
96. P.G. De Gennes, P.A. Pincus, Pair correlation in a ferromagnetic colloid. Phys. Kond. Mat. **11**, 189–198 (1970)
97. N.K. Myshkin, L.V. Markova, Y.E. Kirpichenko, Field optic-magnetic detector application to used oil analysis. in *Proceedings of the 11th International Colloquium of Industrial and Automotive Lubrication*, Esslingen, vol. III, pp. 1975–1978, 1998
98. J.S. Stecki, B.T. Kuhnell, *Ferrographic analysis of high speed gears operating under steady and variable load conditions* (Proceedings of International conference on Condition Monitoring, Brighton, England, 1986), pp. 85–104
99. D.L. Naas, Pat. US5402113, G08 B017/10. Metal particle detector apparatus for non-conducting fluid systems—No.08/112,936. Filed 30.08.93. Published 28 Mar 95
100. British Coal Wear Particle Atlas (Headquarters Technical Department, 1984), p. 14
101. Swansea Tribology Center, *A Guide to Wear Particle Recognition for Use with the Rotary Particle Depositor* (Swansea Tribology Center, UK, 1990), p. 40
102. C.M. Stellman, K.J. Ewing, F. Bucholtz, I.D. Aggarwal, Monitoring the degradation of a synthetic lubricant oil using infrared absorption, fluorescence emission and multivariate analysis: a feasibility study. Lubr. Eng. **56**(3), 42–52 (1999)
103. G.F. Hayes, Observation of association in a ferromagnetic colloid. J. Colloid Interface Sci. **52**(2), 239–243 (1975)

Chapter 6
Trends in On-line Tribodiagnostics

6.1 Introduction

Increase in reliability of complex machinery can be achieved by the following ways: the development of integrated diagnostic systems, the creation of intelligent (SMART) systems, and the development of software including artificial intelligence. Sensors can be both built directly into the machine lubrication system and used as portable instruments in field conditions and industrial laboratories. Fast progress in microprocessors will provide the realization of justified criteria for decision-making on the tribosystem condition. The development of modern electronic technologies promotes the improvement of the accuracy and reliability of new diagnostic systems.

6.2 Modular Systems

Nowadays, modular (integrated) diagnostic systems giving an improved reliability of condition monitoring are becoming more and more popular. Such systems can combine the sensors measuring oil characteristics (IR-sensors, ferrographs, particle counters, etc.) with vibration analysis sensors that improves considerably the validity of data provided by the system [1, 2].

Modular systems of analysis can provide a more complete online estimate of the condition and contamination of oil and the wear debris concentration. Depending on their configurations they can determine the oil purity and oxidation number, data on the number of ferromagnetic and non-ferromagnetic particles, and their size distribution. Sensors can be both built directly into the machine lubrication system and used as portable instruments in field conditions and industrial laboratories. In some cases the diagnostics involving the determination of standard oil characteristics is needed to evaluate the condition of complex and expensive equipment.

N.K. Myshkin and L.V. Markova, *On-line Condition Monitoring in Industrial Lubrication and Tribology*, Applied Condition Monitoring 8,
DOI 10.1007/978-3-319-61134-1_6

Thus, in US patent [2] a modular system is proposed for determining oil characteristics similar to standard ones measured in research laboratories. The system includes a viscosity sensor, an X-ray fluorescent spectrometer (X-ray fluorimeter), and an IR-spectrophotometer (Fig. 6.1). The test oil is supplied to the system from the main line of the equipment to be diagnosed, and before the X-ray fluorimeter it is cooled. The characteristics estimating both the condition of oil and that of the machine as a whole are evaluated. Thus, viscosity variation shows the possible presence of fuel or water in oil or the deterioration of its properties caused by overheating and oxidation. Changes in oil properties caused by chemical reactions are determined by IR spectral analysis that can be also used to find total acid and alkaline numbers of oil. The use of the IR-spectrophotometer also yields data on the water content in oil. The leakage of coolant into the lubrication system is detected by the X-ray fluorimeter when boron, chromium or some other elements such as iodine and strontium being the coolant components appear in oil. System monitoring and data collection are carried out by the controller that comprises

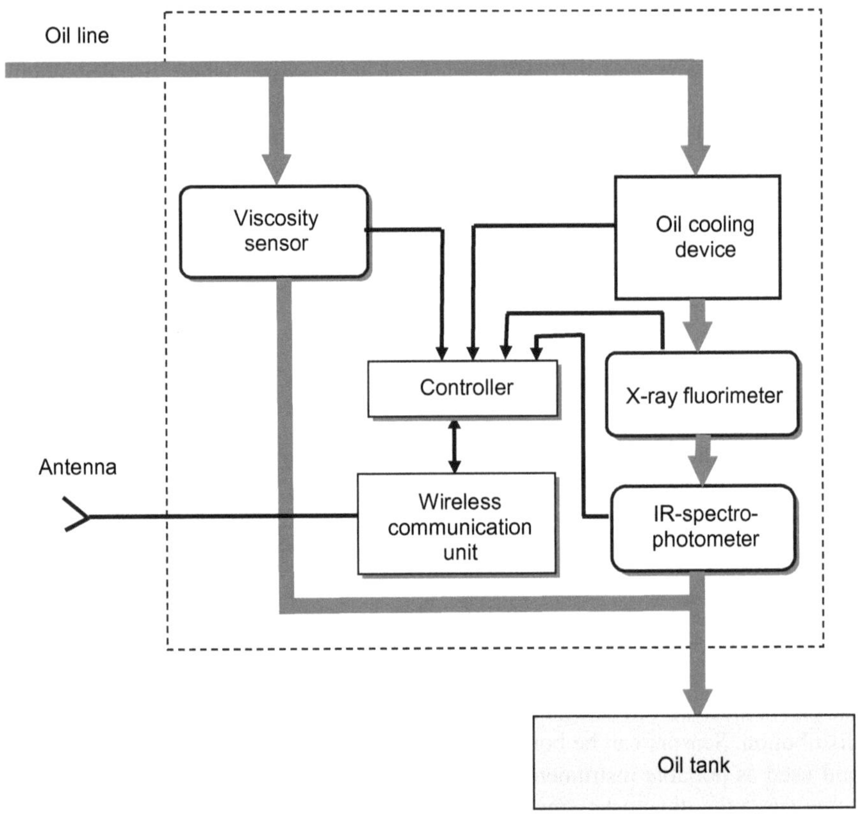

Fig. 6.1 Modular system for determination of standard oil characteristics

microprocessor, memory chip, digital inputs/outputs, analogue input, and the storage device for standard characteristics. Wireless data transmission is performed by the unit with an antenna. Continuous registration produces the database that can be used to predict wear and the schedule of current repairs.

Such modular system gives a valid conclusion on the equipment condition. However, because of its complexity and high cost its usage is rather restricted and can be justified only when applied at plants for the diagnostics of major equipment.

To monitor the tribosystems in real time low cost and miniature sensors are developed [3–5].

Miniature low-cost integrated in-line oil monitoring detector is described elsewhere [5–7]. The detector consists of an integrated probe and electronic module (Fig. 6.2). The integrated probe with three sensors enables in-line monitoring of the lubricating oil in tribosystem by measuring and evaluating four diagnostic parameters simultaneously: chemical degradation, total contamination, water content, and temperature of the oil. This set of parameters allows monitoring of the performance of oil of turbines, compressors, hydraulic system, etc., in real time.

Figure 6.3 shows the design of the integrated probe. The sensor of chemical destruction and total oil contamination comprises the light emitting source *1*, measuring photodetector *2*, feedback photodetector *5*, and optical windows *3* and *4*. When the sensor is installed in the tank, the oil fills the gap between the optical windows. The light from source *1* passes the oil layer *t* thick and illuminates the

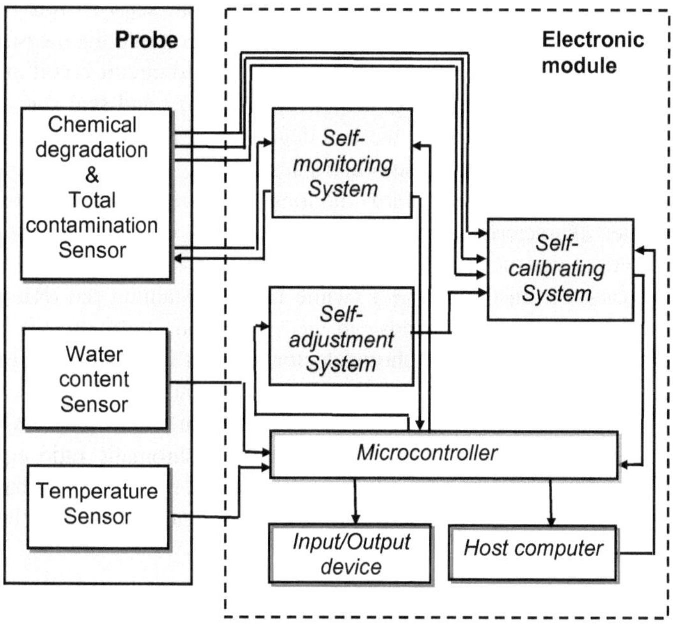

Fig. 6.2 Block diagram of an integrated detector

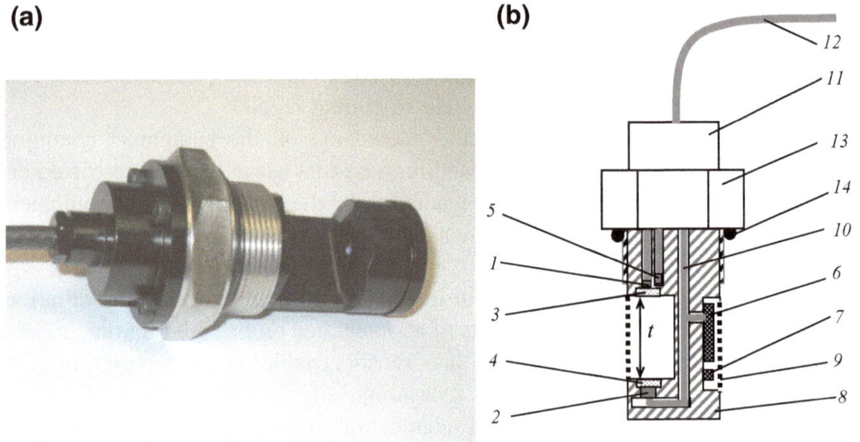

Fig. 6.3 Integrated probe: **a** general view; **b** probe design; *1* light emitting source; *2* measuring photodetector; *3*, *4* optical windows; *5* feedback photodetector; *6* water content sensor; *7* temperature sensor; *8* housing; *9* protective mesh; *10* wires; *11* cover; *12* electrical cable; *13* nut; *14* seal ring

sensitive area of the measuring photodetector *2*. The feedback photodetector *5* serves to stabilize the light emission of source *1*. Sensor *6* of water content and temperature sensor *7* are glued in the slot of housing *8* with epoxy resin. The housing is enclosed in the protective mesh *9* to protect the sensors from mechanical damage and prevent penetration of air bubbles from the oil into the measuring zone. Wires *10* connect the sensors with a preamplifier printed circuit board under cover *11*. The probe is mounted in the oil tank with the nut *13* and seal ring *14*.

To assess the oil oxidation and thermal degradation, the method was developed while estimating the chemical destruction parameter Δ_{CR} by the color shift of the used oil compared to fresh oil towards the long wavelength of the spectrum [6, 8]. This parameter characterizes the alteration of the oil chemical structure due to oxidation, thermal destruction, decomposition of additives, etc. The polychromatic emission is passed from the source 1 (white LED) containing red (R), green (G), and blue (B) wavelength bands in its spectrum through the oil. The three-component color (R, G, B) photodetector 2 serves to measure the intensity of the light emission that passes through the oil simultaneously in three spectral bands: red ($\Delta\lambda_{\mathrm{R}}$ = 590–750 nm), green ($\Delta\lambda_{\mathrm{G}}$ = 490–610 nm), and blue ($\Delta\lambda_{\mathrm{B}}$ = 400–510 nm). The oil color shift is determined from the chromatic ratio equal to the relation between the signals of the photodetector in the red and green bands. Then the chemical destruction diagnostic parameter Δ_{CR} is determined from the formula:

$$\Delta_{CR} = CR_{\mathrm{used}} - \mathrm{CR}_{\mathrm{fresh}}, \tag{6.1}$$

where CR_{fresh} is the chromatic ratio of the fresh oil, $CR_{\mathrm{fresh}} = U_{\mathrm{R,fresh}}/U_{G,fresh}$;

CR_{used} is the chromatic ratio of the used oil, $CR_{\mathrm{used}} = U_{R,\mathrm{used}}/U_{G,used}$;

$U_{R,\text{fresh}}$, $U_{R,used}$ are the photodetector output signals in the red wavelength band during analysis of the fresh and used oils, respectively; and $U_{G,fresh}$, $U_{G,\text{used}}$ are the output signals in the green band during analysis of the fresh and used oils, respectively.

The chemical destruction parameter as a relative value is independent on temperature and fluctuations of the source's light intensity, providing a high sensitivity of monitoring of the oil condition.

At the same time the detector allows estimating the total oil contamination by impurities from the environment (sand, dust, metallic particles, etc.) or impurities that appear during tribosystem operation (corrosion products, wear debris). The indices of total oil contamination are defined by the changes of the optical density of the used oil versus the fresh oil in three spectral bands—red (ΔD_R), green (ΔD_G), and blue (ΔD_B)—from the formulas:

$$\Delta D_R = D_{R,\text{used}} - D_{R,\text{fresh}} = \ln(U_{R,\text{fresh}}/U_{R,used}); \tag{6.2}$$

$$\Delta D_G = D_{\text{G,used}} - D_{\text{G,fresh}} = \ln\left(U_{G,\text{fresh}}/U_{G,used}\right); \tag{6.3}$$

$$\Delta D_B = D_{B,\text{used}} - D_{B,\text{fresh}} = \ln\left(U_{B,\text{fresh}}/U_{B,used}\right), \tag{6.4}$$

where $D_{R,\text{fresh}}$, $D_{G,fresh}$, $D_{B,fresh}$ are the optical densities of the fresh oil in the red, green, and blue bands, respectively; and $D_{R,used}$, $D_{G,used}$, $D_{B,used}$ are the optical densities of the used oil in the red, green and blue bands, respectively.

Water content in the oil and oil temperature are estimated by commercial sensors.

A modified oil moisture content sensor HIH-3610 (Honeywell Inc.) is used to measure air humidity. The sensor is a capacitor inserted between the absorbing layers of a thermosetting polymer with dielectric permeability varying in response to the quantity of absorbed water. The sensor is able to determine the relative saturation (*RS*) of the oil by water. Temperature sensor *7* TC1047 (Microchip Technology Inc.) measures oil temperature within the range −40 to +125 °C.

The electronic module of the detector comprises self-calibration, self-adjustment and self-monitoring circuits controlled by a microcontroller. The memory of the microcontroller stores the initial parameters of the fresh oil and the limit values. Microprocessor reads and processes the signals of the sensors, compares the obtained values with the thresholds and exchange data with the I/O device and host computer.

Laboratory test of the detector. The detector was tested with oil samples after their artificial aging and with oil sampled from operating tribosystems. Samples of both mineral and synthetic oils were analyzed. The diagnostic parameters of the detector and the standard parameters, such as viscosity and total acid number (TAN) of the oils, were compared.

Assessment of the dependence of diagnostic parameters on the concentration of mechanical impurities in oil. Sources of mechanical impurities can be both debris and particles that penetrate into the oil from the environment (sand, dust, etc.).

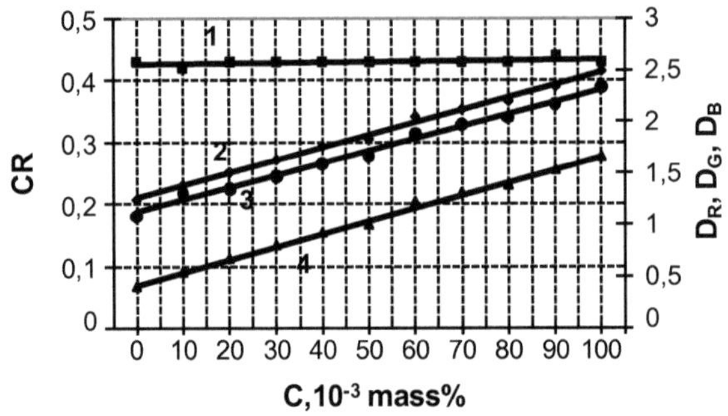

Fig. 6.4 Dependences of diagnostic parameters on concentration of mechanical impurities in oil: *1* chromatic ratio; *2–4* indices D_R, D_B, D_G, respectively

A standard contaminant (carbonyl iron particles with mean size 7.6 μm) was used to simulate contamination. The particles were introduced into fresh T100 mineral oil at concentration within (0–100) × 10^{-3}mass% with the increment 10 × 10^{-3}mass%. The results of measurements in Fig. 6.4 show that the chromatic ratio *CR* remains unchanged when the concentration of mechanical impurities grows; i.e., the chemical destruction parameter Δ_{CR} does not depend on the concentration of the mechanical impurities and it is equal to zero, proving that the chemical structure of the oil does not change. Meanwhile, the indices of total oil contamination ΔD_R, ΔD_G, ΔD_B increase linearly with the content of particles providing valid information on the oil contamination.

Analysis of mineral oils. During the tests, T100 mineral oil was aged following the standard method ASTM D2070. In the process of aging, the oil was sampled every 24 h and analyzed. During thermal ageing, the spectra of transmission of the oil samples obtained with an Opron-3000 spectrometer are placed into the long-wave (red) band (Fig. 6.5a); the oil chemical destruction index Δ_{CR} increases correspondingly (Fig. 6.5b, curve 4). Figure 6.5c shows the correlation between the chemical destruction index and the oil viscosity. The indices of total oil contamination (Fig. 6.5b) also increase due to changes in the oil's chemical structure and its optical transmittance (Fig. 6.5a).

Besides, samples of T100 and highly viscous AW220 mineral oils from the hydraulic systems were analyzed. Figure 6.6 shows that the chromatic ratio and viscosity of the oils increases as their chemical structure changes in the operation of the hydraulic systems.

Analysis of synthetic oils. During studies of changes in the diagnostic parameters of synthetic oils, only samples from operating tribosystems were tested, because the artificial thermal degradation of such oils in laboratory conditions would consume a lot of time.

Samples of SHD-1 (based on polyalphaolefins) and SHD-2 (based on polyglycoles) oils were taken from operating hydraulic systems after equal intervals. Changes in the chromatic ratio of the used oil compared to the fresh one and larger total contamination in operation of the system reflects gradual impairment of the performance of the oils: changes in the chemical structure and increases in total contamination correlate with higher viscosity (Fig. 6.7a, b). Increases in the

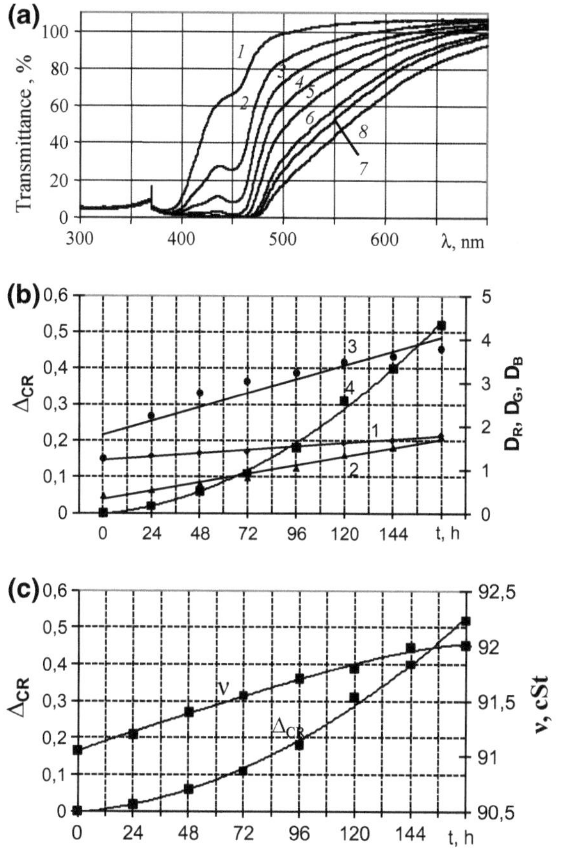

Fig. 6.5 Results of analysis of T100 mineral oil during artificial thermal aging: **a** changes in oil transmission spectrum in visible waveband in response to increased aging duration (*1*—0 h; *2*—24 h; *3*—48 h; *4*—72 h; *5*—96 h; *6*—120 h; *7*—144 h; *8*—168 h); **b** changes in diagnostic parameters of the sensor (*1–3* indices D_R, D_B, D_G, respectively; *4* chemical destruction index; *c* correlation between changes of chemical destruction parameter (*1*) and oil viscosity (*2*)

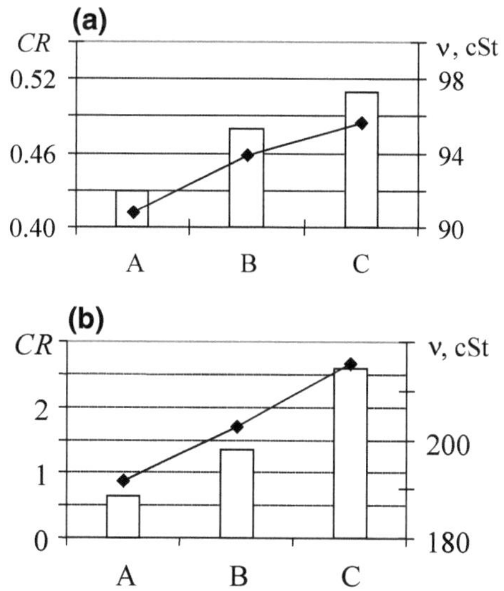

Fig. 6.6 Changes in chromatic ratio (diagram) and kinematic viscosity (curves) of T100 **a** and AW220 **b** mineral oils at their operation in hydraulic systems: *A* fresh oil sample; *B*, *C* used oil samples; operation time of sample C is longer than sample B

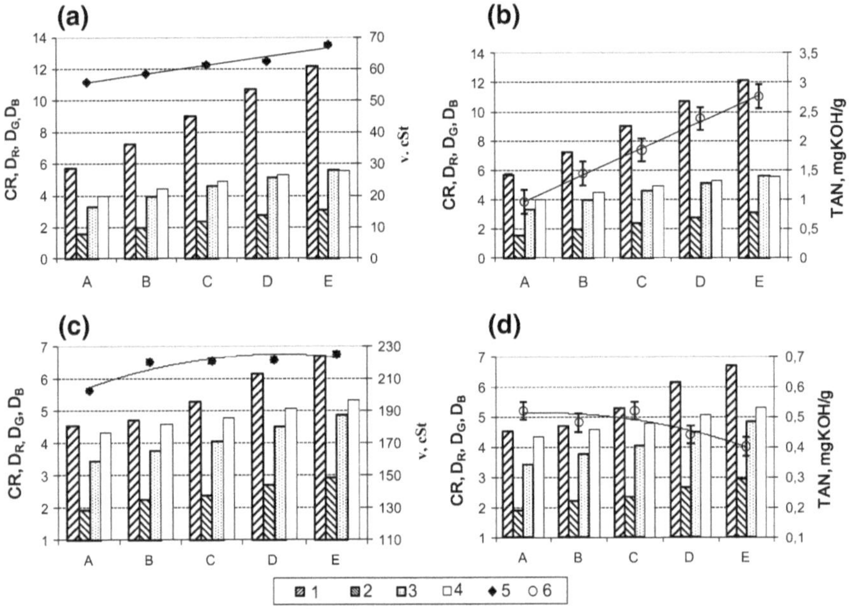

Fig. 6.7 Correlations of chromatic ratio (*1*) and total contamination indices D_R, D_B, D_G, (*2*–*4*) of SHD-1 (a, b) and SHD-2 (c, d) of synthetic oils after operation in hydraulic systems with their kinematic viscosity (*5*), and total acid number (*6*): sample A—fresh oil; B, C, D, E—samples after longer time of the oil operation

chromatic ratio and viscosity are accompanied by an increase in the total acid number of the SHD-1 oil, i.e., the extent of oxidation (Fig. 6.7b). If the acid number of the SHD-2 oil drops slightly during operation (Fig. 6.7d), it may results from the fact that some anticorrosion and antiwear additives being acidic deplete in operation, so that the total oil acidity can reduce while the base of the oil oxidizes.

Test of water content and temperature sensors. The water concentration sensor and the temperature sensor were tested in a tailored setup with the integrated detector, platinum Pt100 thermometer (measurement error ± 0.2%), WS04 oil humidity sensor (Pall Corporation, measurement error ± 2%), heater, stirrer, and water bath, which is used to increase the air humidity over the oil surface and the oil moisture concentration. The container was filled up with Rando-HD-32 fresh hydraulic oil. During the experiment, the readings of the integrated sensor, humidity sensor, and platinum thermometer submersed in the oil were registered. The obtained results show a high correlation of the readings of the temperature sensor and the platinum thermometer, as well as the readings of the water concentration sensor and the humidity sensor of the integrated sensor (Fig. 6.8).

The validity of the detector output data is confirmed, which assures its broad application for in-line assessment of the performance of the oil and the tribosystem condition providing the high information content, and the level of automation. In particular, the detector can be applied to diagnose hydraulic systems, compressors,

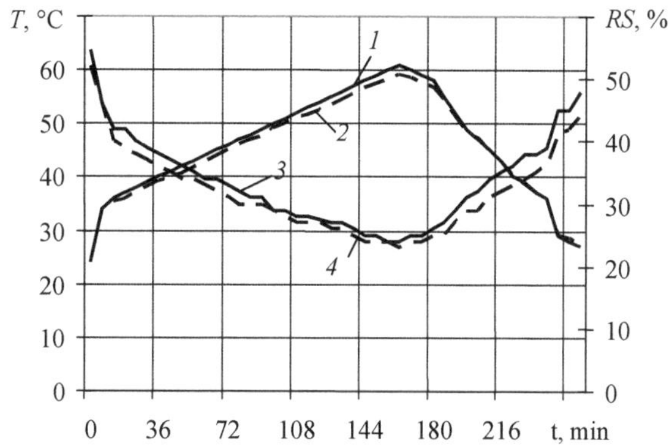

Fig. 6.8 Changes in readings of temperature sensor (*1*), platinum thermometer (*2*), water content sensor (*3*), and WS04 humidity sensor (*4*) in process of tests

turbines, transformers, and other machines for continuous *in situ* monitoring multiple parameters of oil performance for predicting optimal oil change intervals, and equipment condition control.

The development of modern technologies gives rise to the application of modular systems with relatively cheap sensors produced by the technology similar to the integrated one. They are a set of sensing elements deposited to the common substrate [9, 10] and can be built in the oil line of machines. One example is the modular system (Fig. 6.9a) [10] with the oil analyzer produced by integrated technology. It comprises silicon substrate *1* with sensors of acidity *2*, chemical composition *3*, conductivity *4*, and temperature *5* of oil (Fig. 6.9b). Acidity sensor *2* consists of reference electrode *2′* made of silver chloride (AgCl) and pH-electrode *2″* made of palladium oxide (PdO). Chemical sensor *3* is used to determine the water content and/or oil oxidation and comprises three electrodes one of which is silver working electrode *3′*, the second one is reference electrode *3″* made of silver chloride, and the third one is silver-plated measuring electrode *3‴*. Alternating or direct voltage is applied to working electrode *3′* and current passes between electrodes *3″* and *3‴*; the current characteristics vary as a result of electrochemical processes running on the working electrode surface. Conductivity sensor *4* consists of two gold electrodes *4′* and *4″*. The content of metal wear particles and water is determined by the oil conductivity. Oil temperature sensor *5* is a gold track of a certain length and width. The conductivity of the track varies with changing the oil temperature. The sensors have contact pads *6* to be connected to the analogue signal processor (Fig. 6.9a) where output signals are amplified and transmitted to the controller analogue input. It processes digital signals and interprets the data. The results are shown at the display and transmitted to the central processor through the wireless communication unit that comprises a transmitter, a receiver, and an antenna. Wireless communication reduces costs, noises, and other problems arising when using the cable communication.

Technology of oil analyzer manufacturing allows for achieving small dimensions and mass of the instrument as well as its low cost. However, reliability and data validity of modular sensors produced by integrated technology are still low [11].

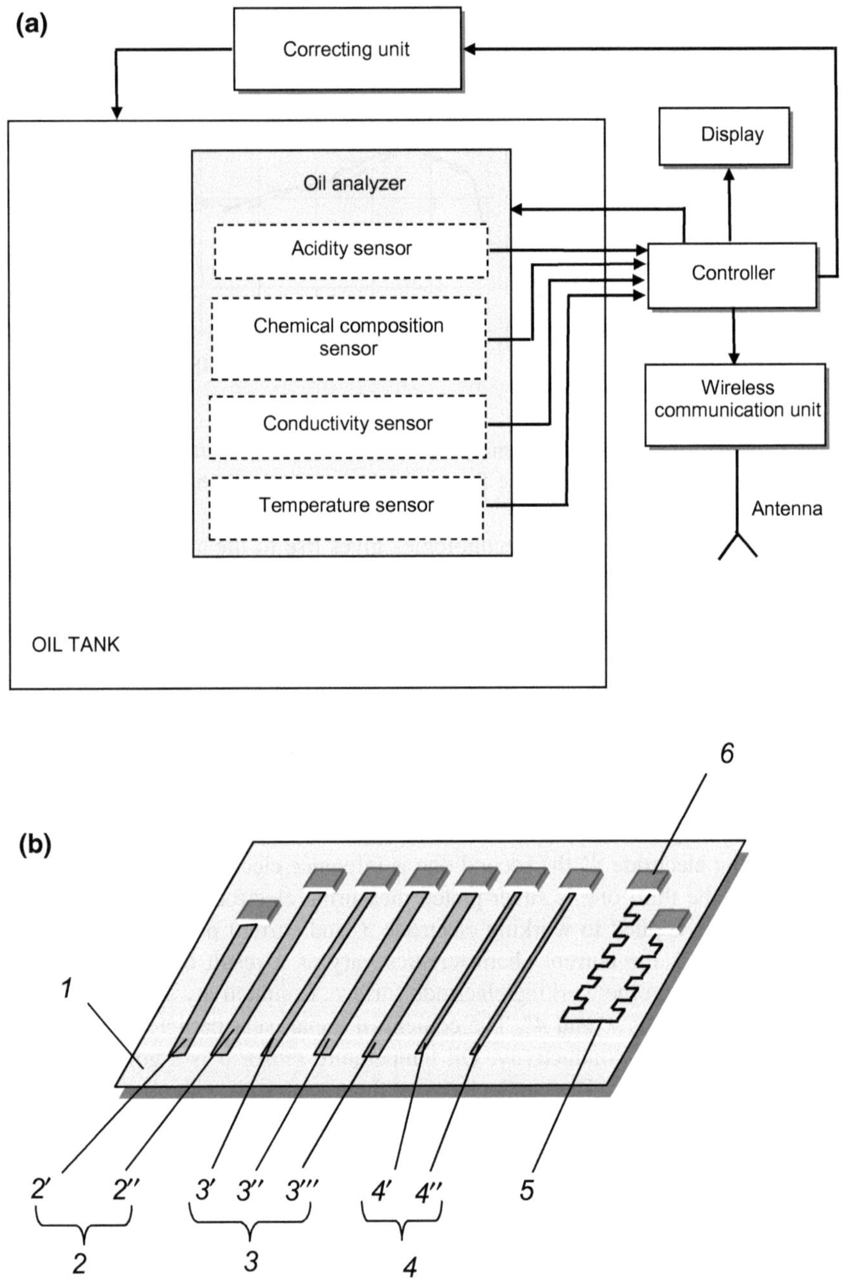

Fig. 6.9 Modular system with oil analyzer built in oil line: **a** system diagram; **b** oil analyzer produced by integrated technology (designations are given in the text), [10]

6.3 Intelligent (SMART) System

When improving the reliability and operational safety of complex equipment the concept of intelligent or SMART (Self-Monitoring, Analysis, and Reporting Technology) systems becomes widely applied. Such systems monitor continuously the equipment condition and wear mode, predict the tribosystem life and inform the staff on the necessity of maintenance [12, 13].

Let us consider the diagnostic SMART system using vehicles as an example [14]. The complete system (Fig. 6.10) includes an on-board system mounted on vehicle and central diagnostic system outside it. The on-board system estimates the condition of vehicle parts and lubricants and transmits data to the central system where the obtained and archive data of the particular vehicle and the standard data are used to determine the actual condition of the mechanism being diagnosed.

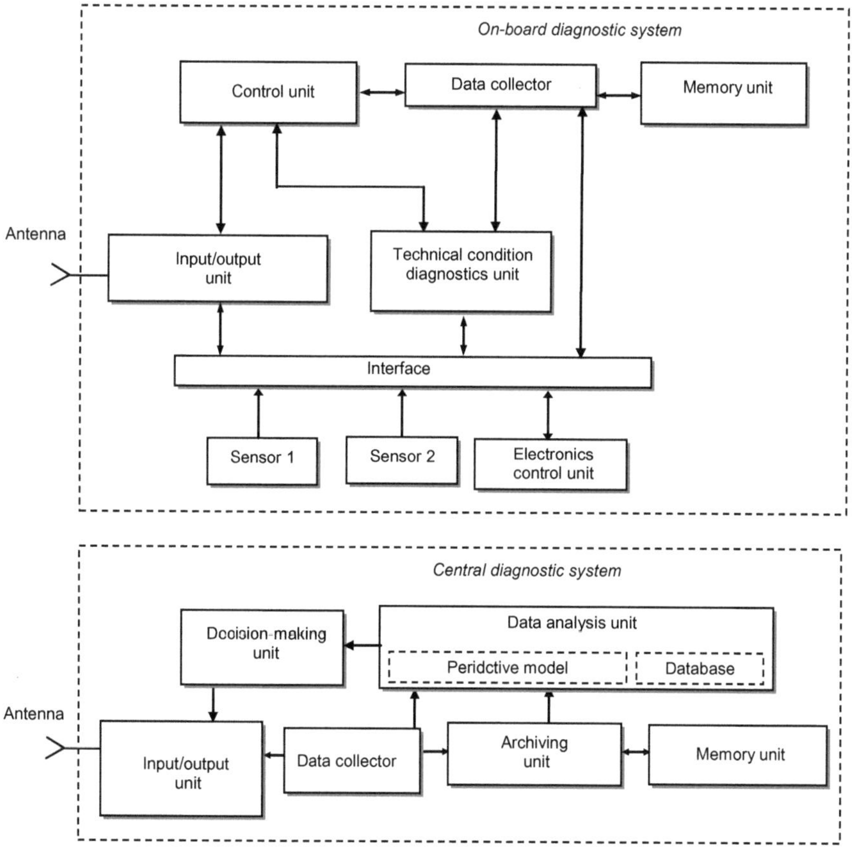

Fig. 6.10 Diagram of diagnostic SMART system of vehicle

The operation of the on-board monitoring system is based on the continuous collection of data from the sensors and their processing to get information on the object technical condition. Information from the sensors set in the car is transmitted to the data collector through an interface, then to storage and diagnostic units for analysis with consideration of the model and database. The analysis result is transmitted to an input/output unit through an interface and then to the central diagnostic system. The on-board system is controlled by a control unit.

The central system gets data from the on-board system to the data collector. Data from service centers also come here and transmitted to the analysis unit and to storage device through an archiving unit. They are analyzed in order to determine the necessity of preventive maintenance and to search and correct defects. Data analysis in the diagnostic system uses predictive models. Initially a general model is developed based on the data obtained from one-type vehicles. Then it is specified and adjusted to a particular vehicle using data obtained in its testing. When the necessity of maintenance is determined the information is sent to the driver and to the service station. It contains data on the vehicle and required maintenance actions.

SMART systems are now being developed that are capable of performing corrective actions to maintain equipment in working condition. For example, an oil analyzer [10] may include the correcting unit that can perform corrective actions such as the introduction of pure oil, oxidation inhibitor or viscosity improver.

The major component of any intelligent diagnostic system is a modular system of sensors that have to meet requirements imposed to SMART sensors, which combine a sensing element with a microcontroller and perform the following functions [15]:

- process information;
- transmit and receive signals in real time by network hardware which can be a cable, telephone line or wireless communication;
- perform self-diagnostics and self-calibration using the storage data.

The SMART sensors described above (Figs. 6.1 and 6.10) comprise controllers performing the processing and preliminary analysis of measured data, data readout, and data input and output through the wireless communication unit.

6.4 Artificial Intelligence Methods of Data Processing

The development of modular measuring systems requires reliable systems for processing and analysis of data and setting a valid diagnosis. In particular, the technology of data fusion is applied, i.e. the combination of data from different sensors. Each sensor yields a fragment of data on the oil condition that consists of the dependent and independent components. Independent components are used to fill further information space while dependent ones are used in combination (fusion). Data fusion is the algorithmic processing of sensor data in order to compensate considerable data fragmentation since the tribosystem condition cannot be determined reliably based on just one characteristic.

Artificial intelligence systems have contributed recently into the development of diagnostic systems and sensors, neural networks, and expert systems [2, 13, 14].

The algorithm of neural networks widens its use fusing the data coming from many sensors consisting of simple computing elements (neurons). It gets a set of input signals and outputs the respective response providing the problem solution. The neural network contains internal characteristics of neurons and the whole structure, and varying them it can learn. At this stage a set of reference situations with known solutions is specified and the neural network finds relationships between input signals and solutions. It is capable of generalizing the acquired skill of problem solution and giving a prognosis for new input signals not included at the teaching stage. This allows for developing an integrated multi-dimensional model of the tribosystem in which the neural network uses complex relationships among variables for self-learning. Neural network training includes the acquisition of input and output data on oil samples comprising indicators similar to those under investigation, i.e. information on oil characteristics under determination. The trained neural network is capable of giving a correct output (tribosystem condition) even when it has never before met the same set of data.

The implementation of neural networks requires a large amount of preliminary information. Expert systems have been used to solve the problems with incomplete and fuzzy initial information, as well as the fuzziness of objectives. Systems based on fuzzy logic with linguistic (verbal) variables hold a special place among them. In other words, a transition is made from numerical to verbal data and from a verbal qualitative description of an object, which is typical of human thinking, to numerical quantitative estimates of the state of an object. An advantage of fuzzy logic is the possibility of using the expert knowledge of the object in the form of linguistic statements, such as "if then". Expert systems with fuzzy logic have been used for monitoring the state of oil-filled systems.

Thus, the monitoring of furane content in the transformer oil [16] and the diagnostics of tribosystem wear based on an analysis of lubricating oil [17] were developed using the fuzzy logic. The input parameters of the decision-making system constructed using the fuzzy logic can include the set of values of diagnostic parameters of oil-quality sensors.

The fuzzy logic method application for decision making on the lubricating oil state.

The fuzzy logic method based on the Mamdani algorithm for decision making on the lubricating oil state is implemented in the device, including the information-collection and decision-making units [18]. In general, the information-collection unit can represent any set of sensors that reflect the state of the system. The modular in-line detector (Fig. 6.3) and magnetoelastic viscometer [19] were used to estimate the lubricating oil performance. The device can be built into the lubrication system for real-time diagnostics or used for laboratory analysis of the oil samples. The structural diagram of the method is shown in Fig. 6.11.

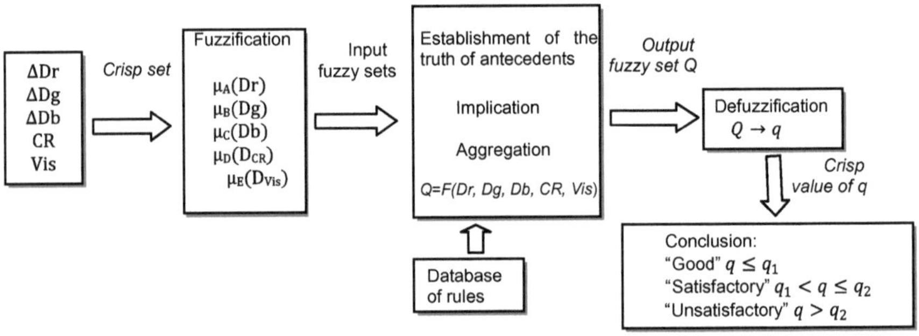

Fig. 6.11 Structural diagram of the fuzzy logic method for solution of the decision-making problem on the state of lubricating oil based on the Mamdani algorithm

The input parameters are the indices of total oil contamination measured in the red (ΔD_R), green (ΔD_G), and blue (ΔD_B) ranges and chromatic ratio (CR) measured by integrated detector and kinematic viscosity (Vis) measured by the magnetoelastic viscometer.

The values of information parameters from the data collection system are supplied to the decision making system. The decision-making unit on the lubricating oil state implements the fuzzy logic method based on the Mamdani algorithm [20].

One of the following values of the output linguistic set $Q = \{Q_1, Q_2, Q_3\}$, which characterizes the oil state is chosen based on the analysis of the input variables: Q_1, meaning "good"; Q_2, meaning "satisfactory"; or Q_3, meaning "bad." The algorithm of forming the logical conclusion based on the analysis of the input parameters includes the following procedures.

1. *Formation of a database of rules* for input variables influence on the value of output one; these rules represent a set of logic statements:

   (1) If the optical density *Dr* is normal, *Dg* is normal, and *Db* is normal, the chemical destruction factor *CR* is normal, and the oil viscosity *Vis* is normal, then the state of the oil is good (Q_1). The weighting coefficient of the rule, which takes into account the degree of influence of this rule on the decision-making, $W = 1$.

   (2) If *Dr* is increased, *Dg* is increased, *Db* is increased, *CR* is increased, and *Vis* is increased, then the oil state is satisfactory (Q_2). The weighting coefficient of the rule $W = 1$.

   (3) If *Vis* is high, then the oil state is bad (Q_3). The weighting coefficient of the rule $W = 0.5$.

   (4) If *Dr* is high, then *Dg* is high, *Db* is high, and *CR* is high, and the oil state is bad (Q_3). The weighting coefficient of the rule $W = 1$.

2. *Fuzzification* (introduction of fuzziness) of inputs. The input variables *Dr, Dg, Db, CR, and Vis* are a crisp set, i.e., the set of some numbers. At the fuzzification stage these values are transformed into the fuzzy format, i.e., the correspondence

between the numerical value of the input variable and the value of the input linguistic variable (fuzzy set $X = \{X_1, X_2, X_3, \ldots\}$) is established. In the general case, a fuzzy set is established for each variable.

The fuzzy set $A = \{A_1, A_2, A_3\}$, where the values of the linguistic variable are A_1, meaning "normal'; A_2, meaning "increased"; and A_3, "high" corresponds to variable *Dr*; set *B* corresponds to value *Dg*; set *C* to *Db*; set *D* to *CR,* and set *E* to *Vis.*

The correspondence between the input variable and the fuzzy set is determined using the membership function $\mu_{Aj}(Dr), \mu_{Bj}(Dg), \mu_{Cj}(Db), \mu_{Dj}(CR), \mu_{Ej}(Vis)$, which takes values in the range [0, 1] and characterizes the degree of correspondence of the input numerical variable x_i to the linguistic definition. The fuzzification of all input variables is performed using the Gaussian membership functions of the form (6.5)

$$\mu_{X_j}(x_i) = e^{-\frac{(x_i - c)^2}{2\sigma^2}}, \tag{6.5}$$

where σ is the mean square deviation of the function and *c* is the mathematical deviation (determines the position of the function maximum). Each value of the input linguistic (fuzzy) set *Xj* normal, increased, and high corresponds to particular ranges of numerical values of the input variables *xi* = [*Dr*, *Dg*, *Db*, *CR*, and *Vis*].

3. *Finding the degree of truth of the antecedents.* In the considered case, the input of the logic operator are the numerical values of the input membership functions $\overline{\mu}_{Aj}(Dr), \overline{\mu}_{Bj}(Dg), \overline{\mu}_{Cj}(Db), \overline{\mu}_{Dj}(CR), \overline{\mu}_{Ej}(Vis)$ which correspond to the measured values of *Dr, Dg, Db, CR,* and *Vis.* To obtain the quantitative value of truth (cutoff level) for all antecedents (the "if" part of the rule) of a particular kth rule, logic operator AND is used; it is implemented by the operator of finding the minimum value of the membership function: $\overline{\mu}_k = min\{\overline{\mu}_{Aj}(Dr), \overline{\mu}_{Bj}(Dg), \overline{\mu}_{Cj}(Db), \overline{\mu}_{Dj}(CR), \overline{\mu}_{Ej}(Vis)\}$.
4. An *implication* is the procedure of finding the consequent ("then" part of the rule), i.e., finding the logic conclusion (fuzzy set) of each rule. This set is represented by the output membership function, which corresponds to the value of the linguistic variable.

Before the implication, the weighting coefficients ($W \leq 1$) of the rules are determined, which define the degree of influence of a particular rule on the decision-making. The degree of truth of each of the consequents is equal to the algebraic product of the corresponding value and the weighting coefficient W_k: $\overline{\mu}'_k = \overline{\mu}_k \cdot W_k$.

The values of *Wk* are given in the rules above.

The input for the implication procedure consists of the numerical values $\overline{\mu}'_k$ (cutoff levels) of each rule, and the output consists of the values of the fuzzy set

$\mu_k(q)$. The implication is performed for each rule based on the logic operation AND, which cuts off the output fuzzy set on the level $\overline{\mu}'_k$

$$\mu_k(q) = min\{\overline{\mu}'_k, \mu_{Qj}(q)\},$$

where $\mu_{Qj}(q)$ is the membership function of the linguistic value (good, satisfactory, bad) of output variable q defined on the set Qj (oil state). The membership function $\mu_{Qj(q)}$ of the set Q_j is taken in the form of the triangular function

$$\mu_{Qj}(q) = \begin{cases} 0, & q < a_j \\ \frac{q-a_j}{b_j-a_j}, & a_j \le q \le b_j \\ \frac{q-a_j}{b_j-a_j}, & b_j \le q \le c_j \\ 0, & q > c_j \end{cases}, \tag{6.6}$$

where a_j and c_j characterize the base of the triangle, while b_j is taken in its vertex.

5. *Aggregation* is the procedure of uniting the membership functions of the output variable. The input of the aggregation process is the set of cutoff output membership functions ($\mu_1(q)$, $\mu_2(q)$, $\mu_3(q)$, $\mu_4(q)$) obtained in the course of the implication for each rule, and the output is the set of the function $\mu_\Sigma(q)$ of the united set sum The aggregation procedure is performed using the method of finding the maximum value as follows:

$$\mu_{\sum}(q) = \max(\mu_1(q), \mu_2(q), \mu_3(q)\mu_4(q)).$$

6. *Defuzzification* of the output variables. The fuzzy logic conclusion is transformed into the crisp value (number) using the defuzzification procedure. In the fuzzy set theory, the defuzzification procedure is similar to finding the characteristics of the position (mathematical expectation, mode, median) of random quantities in the probability theory. In the formulated problem, the defuzzification of the fuzzy set is performed using the center of gravity method according to the formula

$$q_0 = \frac{\int_{min}^{max} q \cdot \mu_{\sum}(q)}{\int_{min}^{max} \mu_{\sum}(q)},$$

where q_0 is the result of defuzzification, q is the variable corresponding to the output linguistic variable Q, $\mu_\Sigma(q)$ is the membership function of the fuzzy set corresponding to the output variable q after aggregation, and min and max are the left and right points of the range of the q variation.

The numerical value of the output variable q_0 is equal to the abscissa of the center of gravity of the area bounded by the plot of the membership function.

The obtained numerical parameter q_0 characterizes the oil state and can be used as the integral numerical oil state factor. The oil state is Q_1 "good" if Q_2 "satisfactory" if and Q_3 "bad" if $q_1 < q_0 \leq q_2$.

Thus, the fuzzy output variable Q corresponds to the particular class Qj, i.e., the crisp conclusion is made on the oil state.

The proposed decision-making method on the lubricating oil state was implemented in MATLAB using the editor for fuzzy inference system (FIS Editor).

The intelligent decision-making method was applied for analysis of the state of engine oil Mobil Super 3000 X1 5 W-40 used in the (2.0T) engine of Opel Vectra 2002. Three intervals of values of the parameters *Dr*, *Dg*, *Db*, *CR*, and *Vis* were used for three linguistic variables *Xj* that characterize these intervals of input values as normal, increased, and high. The critical values of the output numerical parameter q_0 for the analyzed oil are as follows: $q_1 = 3$ and $q_2 = 7$, i.e., the oil state is Q_1 "good" if Q_2 "satisfactory" if and Q_3"bad" if $q_0 > 7$.

Table 6.1 gives the ranges of variations in the input variables (outputs of the integral detector and the viscometer) and the defined parameters of the corresponding Gaussian membership functions $\mu_{Xj}(xi)$, see formula (6.5). Figure 6.12a shows the fuzzification of the input variable *Dg*.

The fuzzification of the output variable is performed using the triangular membership function for a given range of the numerical output parameter q of 0–10. The plots of the membership function $\mu_{Qj}(q)$ of the linguistic value (good, satisfactory, and bad) for the output variable q defined on the set Q_j (oil state) are shown in Fig. 6.12b.

The diagram shown in Fig. 6.13 elucidates the decision-making algorithm on the state of Mobil Super 3000 X1 5 W-40 oil for the case when the input numerical variables have the following values: $D_r = 2.69$; $D_g = 2.61$; $D_b = 2.61$; $CR = 0.88$, and $Vis = 14.3$ cSt.

The diagram shows all four decision-making rules and the following operations: determination of the degree of truth (by the minimum value of the membership functions), implication (as a result, the cutoff output membership function $\mu_{Qj}(q)$, are determined) for each rule, and aggregation of these sets into one fuzzy set for which the numerical value of the output variable is determined as a result of defuzzification using the center of gravity method, $q_0 = 4.13$.

Table 6.1 Ranges and parameters of membership functions for input variables

Input variable x_i	Variation range	Parameters of Gaussian membership function					
		"Normal", $\mu_{X1}(x_i)$		"Increased" $\mu_{Q2}(x)$		"High" $\mu_{Q2}(x)$	
		σ	c	σ	c	σ	c
Dr	1.8 … 6	0.6	1.8	0.8	3.29	0.6	6
Dg	1.2 … 6.7	0.93	1.2	0.93	3.95	0.93	6.7
Db	1.3 … 7.2	0.75	1.3	0.97	3.77	1	7.2
CR	0.4 … 2.5	0.38	0.4	0.32	1.06	0.36	2.5
Vis	13 … 17 cSt	0.85	13	0.85	13.62	0.85	17

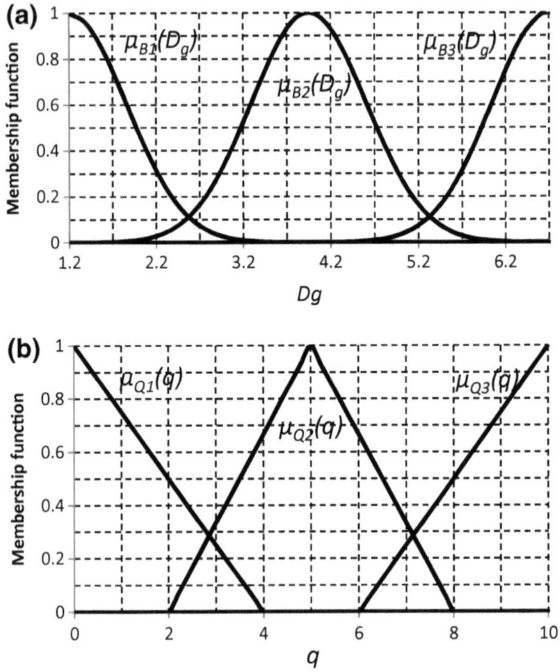

Fig. 6.12 Membership function of **a** input variable Dg in set B, **b** output variable q in set Q

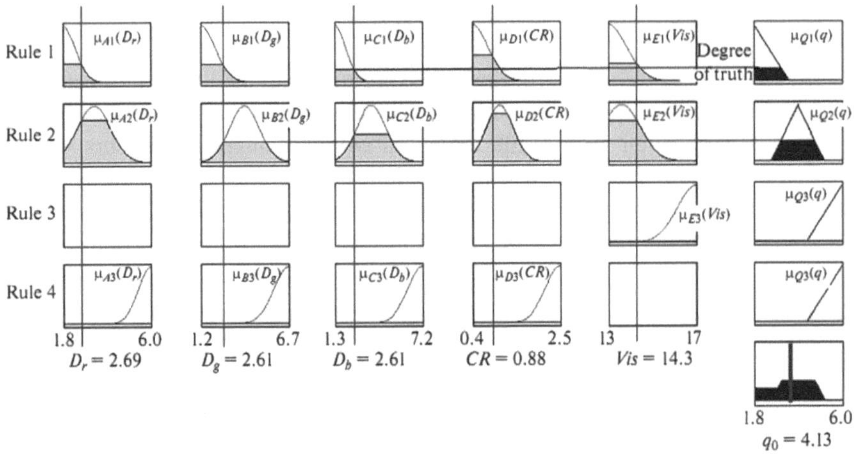

Fig. 6.13 Diagram elucidating the decision-making algorithm on the state of Mobil Super 3000 X1 5 W-40 oil for the following input parameters: D_r = 2.69; D_g = 2.61; D_b = 2.61; CR = 0.882, and Vis = 14.3 cSt

Table 6.2 Diagnostic parameters and output numerical oil state factor for fresh and used oil

Test oil	Diagnostic parameters					Oil state factor q_0
	Dr	Dg	Db	CR	Vis, cSt	
Fresh	2.01	1.37	1.55	0.52	14.3	2.05
Used	5.72	6.42	6.88	2.01	15.9	8.31

The critical values of the output numerical oil state factor q_0 for the analyzed oil are $q_1 = 3$ and $q_2 = 7$, i.e., the oil state corresponds to Q_1 "good" if $q_0 \leq 3$, Q_2 "satisfactory" if and Q_3 "bad" if $q_0 > 7$. Comparing the obtained value of the factor $q_0 = 4.13$ with the critical values, the conclusion is made that the lubricating oil state is satisfactory.

Table 6.2 gives the measured diagnostic parameters obtained in analysis of fresh and used oil after a car run of 107,000 km. The values of the integral factor q_0 determined according to the developed method are 2.05 for fresh oil, which corresponds to the value of the linguistic variable "good", and 8.31 for used oil, i.e., "bad". Thus, it was established that the state of the analyzed used lubricating oil is unsatisfactory and it should be replaced.

The reliability of the operation of artificial intelligence systems is mainly governed by a database on the prehistory of the tribosystem under diagnostics and probable deviations of characteristics to be monitored and their relation to wear modes.

To improve the validity of the diagnosis with artificial intelligence, hybrid technologies and algorithms are now developed combining properties of expert systems and neural networks.

6.5 Conclusions

The development of modern electronic technologies promotes the improvement of the accuracy and reliability of new diagnostic systems. Fast progress in microprocessors will provide the realization of justified criteria for decision-making on the tribosystem condition. This causes the following main trends in tribodiagnostics:

- development of SMART sensors and diagnostic systems which provide the continuous monitoring of tribosystems;
- increasing the number of sensors built as standard elements in the oil line of engines, turbines, and other heavy machinery;
- development of software including fast-evolving methods of artificial intelligence;

- inclusion of tribodiagnostics into the basic program of daily maintenance of equipment;
- reduction of costs of tribodiagnostics methods promoting their more wide application.

Progress in tribodiagnostics will become even more significant when solving the most intricate problems of modern engineering.

References

1. Y. Yin, W. Wang, X. Yan, H. Xiao, C. Wang, An integrated on-line oil analysis method for condition monitoring. Meas. Sci. Technol. **14**, 1973–1977 (2003)
2. B.W. Wilson, T.J. Peters, C.L. Shepard, J.H. Reeves, Pat. US 6561010 G01 N 011/00. Apparatus and method for fluid analysis—No. 09/776,109, Filed 01 Feb 2001. Published 13 May 2003
3. M. Appleby, F.K. Choy, L. Du, J. Zhe, Oil debris and viscosity monitoring using ultrasonic and capacitance/inductance measurements. Lubr. Sci. **25**, 507–524 (2013)
4. B. Von Herzen, S. Van Fleet, H. Fallside, R. Hall, Pat. US 9389215 G01 N 33/28, G01 N 27/02, F01 M 11/10. Multi-modal fluid condition sensor platform and system thereof—No. US 20140130587. Filed 15 May 2014. Published 12 Jul 2016
5. L.V. Markova, N.K. Myshkin, V.M. Makarenko, M.S. Semenyuk, H. Kong, H.-G. Hun, S.V. Ossi, Integral indicator for monitoring the condition of lubricating materials in tribocouples. J. Friction Wear **29**(4), 302–309 (2008)
6. C.V. Ossia, H. Kong, L.V. Markova, Utilization of color change in the condition monitoring of synthetic hydraulic oils. Ind. Lubr. Tribol. **62**(6), 349–355 (2010)
7. H. Kong, H-G Han, L. Markova,V. Makarenko, M. Semenyuk, Pat. US 8155891 G01 N 33/30; G06F 17/40. Integrated in-line oil monitoring apparatus—No. US 20090216464. Filed 27 Aug 09. Published 10 Apr 12
8. H. Kong, E–S Yoon, H–G Han, L. Markova, M. Semenyuk, V. Makarenko, Pat. US 7612874 G01 N 33/28. Method and apparatus for monitoring oil deterioration in real time—No. 11/640,012. Filed 15 Dec 2006. Published. 03 Nov 2009
9. X. Zhu, L. Du, J. Zhe, An integrated lubricant oil conditioning sensor using signal multiplexing. J. Micromech. Microeng. **25**(1), 015006 (2015)
10. F.M. Discenzo, Pat. US 7516650 G01 N 033/30, G01 N 011/00. Lubricity measurement using MEMS sensor—No. US 20070245811. Filed 25 Oct 2007. Published 14 Apr 2009
11. Oil sentinel—Theory of operation. Water contamination and oxidation measurements in base oils & hydraulic fluids, Industrial series sensor system voelker sensors. Inc. Available http://www.vsi-oil.com/Sentinel/OilSentTheoryOper.PDF
12. D.C. Schalcosky, C.S. Byington, Advances in real time oil analysis. Practicing Oil Anal. **11** (2000). Available: http://www.machinerylubrication.com/Read/138/real-time-oil-analysis
13. T.L. Blevins, M.J. Nixon, W.K. Wojsznis, Pat. US 6615090 G05 G 011/01. Diagnostics in a process control system which uses multi-variable control techniques—No. 09/499,446. Filed 07 May 2000. Published 02 Sept 2003
14. C.-N. Fiechter, M.H. Goker, D. Grill, R. Kaufmann, T. Engelhardt, A. Bertsche, Pat. US 6609051 G01 M 017/00, G06 F 015/00. Method and system for condition monitoring of vehicles—No. 09/948,938. Filed 10 Sept 2001. Published 19 Aug 2003
15. S. Vadde, S.V. Kamarthi, S.M. Gupta, Modeling smart sensor integrated manufacturing system. Procedings of SPIE—The International Society for. Opt. Eng. **5263**, 30–37 (2004)
16. A. Abu-Siada, S.P. Lai, S.M. Islam, A novel fuzzy-logic approach for furan estimation in transformer oil. IEEE Trans. Power Deliv. **27**(2), 469–474 (2012)

17. S. Ramezani, M. Yousofi, A fuzzy rule based system for fault diagnosis, using oil analysis results. Int. J. Indust. Engineer. Product. Res. **22**(2), 91–98 (2011)
18. L.V. Markova, Intelligent method for monitoring the state of lubricating oil. J. Friction Wear **37**(4), 308–314 (2016)
19. L.V. Markova, N.K. Myshkin, H. Kong, H.G. Han, On-line acoustic viscometry in oil condition monitoring. Tribol. Int. **44**(9), 963–970 (2011)
20. E.H. Mamdani, S. Assilian, An experiment in linguistic synthesis with a fuzzy logic controller. Int. J. Man Mach. Stud. **7**(1), 1–13 (1975)

Conclusions

The reliability of tribosystems (bearings, guides, gears, seals etc.) depends on their monitoring and predictive maintenance. Failure prevention in machinery is becoming more and more urgent, as the society employs high technologies, machine designs is getting complicated, and the consequences of failures and catastrophes are very expensive (e.g., space and aviation accidents).

Oil condition monitoring often can detect the disorder in normal wearing of the tribosystem much earlier than other techniques (e.g. monitoring of vibrations, accelerations, and temperature).

Up-to-date upgraded requirements to the service characteristics of oils have initiated the development of new methods and instruments for their prompt condition monitoring. Production of new devices includes new technologies and element base, which allow improving the validity of devices and reducing their cost and overall dimensions.

The choice of the diagnosis method and instrument should be governed by the type of tribosystem, its importance and cost. To assess the oil condition in automobile engines small-sized and cheap devices can be used. But in heavy machinery where more complete information on the oil condition is required, it is advisable to use integrated devices including a number of sensors. In this case we can make a valid conclusion on the oil efficiency on the basis of a complex analysis.

An important aspect of ensuring the proper operation of tribosystems is the correct selection of oil, according to viscosity and its variation with temperature, as well as monitoring the oil viscosity during operation. If a change in viscosity is detected, subsequent analysis of the oil can identify the cause of such change. Either an increase or decrease in oil viscosity can result in the disturbance of the bearing capacity of tribosystem.

The review of methods of monitoring the oil viscosity shows that at present, portable viscometers requiring periodic sampling of oil are widely applied; however real-time monitoring is necessary for the reliable operation of machines. Among the devices mounted directly into the oil circulation lines and, particularly, into the lubrication systems of vehicles, acoustic devices are the most promising. Research centers carry out works for the perfection of methods for measuring the viscosity and the design of acoustic devices with the purpose of their wide application for

N.K. Myshkin and L.V. Markova, *On-line Condition Monitoring in Industrial Lubrication and Tribology*, Applied Condition Monitoring 8,
DOI 10.1007/978-3-319-61134-1

real–time diagnostics. In parallel with solid–state sensors based on acoustic waves, magnetoelastic viscosity sensors are currently under development; they operate at lower frequencies. Both piezoelectric and magnetoelastic types of the on-line acoustic viscometers have proved to be promising in the real-time monitoring of lubricating oils.

Water may enter the system from the environment and oil flows circulating in the system can entrap free dropped water. In the friction zone water replaces oil and disturbs the lubrication regime, so the presence of water in machine oil is one of the reasons for severe wear of tribosystem. Monitoring of the water content in industrial oils allows for servicing and prevention of failures of equipment. The most common devices for the on-line real time monitoring of the water content in oils are nowadays based on capacitive sensors and polymer absorbing films. Their main disadvantage is poor measurement reliability when the oil saturation with water exceeds 70%. However, their advantage is in cost-efficiency and simple maintenance. The devices with absorbing elements determine the relative oil saturation with water. Yet, if it is required to evaluate the absolute water content the devices without absorbing elements should be used.

The development of the reliable sensors installed in the equipment is of urgency. The problem consists in producing cheap sensitive sensors for detecting water in oil and providing the high-effective monitoring of the lubrication regime of friction units. Another important task is to provide the predictive maintenance with controlling the level of ware content at a given level providing the long life and reliable operation of the equipment.

The service life of the engine oil is restricted by thermal decomposition, duration of additive effect, and contamination with soot particles formed due to incomplete fuel combustion. Long operation with degraded oil can reduce considerably the life of an engine. Therefore, up-to-date requirements to diesel engine operation and their reliability set the problem of measuring both soot concentration and its dispersity.

Up-to-date requirements set the necessity of development of portable monitoring devices built in the car lubrication system which allow one to obtain the on-line data on the oil contamination by soot.

To realize on-line soot content measurement for vehicles it is required to produce a measuring system which is relatively cheap to be installed to large number of cars and fairly reliable to operate in a diesel engine. Soot contained in oil varies its electrical (conductivity and permittivity) and optical properties (optical density). Therefore, available devices built in the car lubrication system are based on electrical and optical methods.

The fluorescent method allows for monitoring the oil oxidation in equipment operation with high reliability of the on-line diagnostics. The sensor implementing the method is distinguished by simple design, reliability, and low cost. It can be used to monitor oil condition in hydraulic systems, compressors, and turbines.

Monitoring of oil oxidation is very important for power engineering, so the techniques used in condition monitoring of tribosystems can be applied in this area. For example results obtained using the fluorescence detector confirmed that this techniques could be used to analyze the transformer oil oxidation in a simple and effective manner.

Considering the techniques used in wear monitoring it is evident that most devices are based on magnetic methods. It is caused by the fact that magnetic methods, when using in the heavy-loaded units made of ferrous materials, make it possible to identify wear debris among all the solid particles contaminating the lubricant.

The development of built-in devices is carried out basing on optical methods, X-ray fluorescent spectroscopy and nuclear magnetic resonance spectrometry. The efficiency of diagnosis depends greatly on the type and size of wear debris. Mechanisms subjected to monitoring generate wear particles of different size, and one technique can not provide efficiency for all mechanisms. It should be chosen according to the preliminary experimental data on mechanism failure. In other cases several methods should be used.

Efficient monitoring methods and devices were developed in Metal-Polymer Research Institute of National Academy of Sciences of Belarus. In some cases these methods were developed and tested in cooperation with the Korea Institute of Science and Technology.

The development of modern electronic technologies promotes the improvement of the accuracy and reliability of new diagnostic systems. Fast progress in micro-processors will provide the realization of justified criteria for decision-making on the tribosystem condition. This causes the following main trends in tribodiagnostics:

- development of SMART sensors and diagnostic systems which provide the continuous monitoring of tribosystems;
- increasing the number of sensors built as standard elements in the oil line of engines, turbines, and other heavy machinery;
- development of software including fast-evolving methods of artificial intelligence;
- inclusion of tribodiagnostics into the basic program of daily maintenance of equipment;
- reduction of costs of tribodiagnostics methods promoting their more wide application.

Resuming the analysis of current techniques and equipment used in condition monitoring of tribosystems and their predictive maintenance the authors can foresee further rapid development of all the directions of diagnostics and condition monitoring in industrial lubrication and tribology.

FSC
www.fsc.org
MIX
Papier aus verantwortungsvollen Quellen
Paper from responsible sources
FSC® C105338

If you have any concerns about our products,
you can contact us on
ProductSafety@springernature.com

In case Publisher is established outside the EU,
the EU authorized representative is:
Springer Nature Customer Service Center GmbH
Europaplatz 3, 69115 Heidelberg, Germany

Printed by Libri Plureos GmbH
in Hamburg, Germany